28th Annual Saudi-Japan Symposium on Technology in Petroleum Refining and Petrochemicals 2018

Dhahran, Saudi Arabia
12 - 13 November 2018

ISBN: 978-1-5108-7743-6

Printed from e-media with permission by:

Curran Associates, Inc.
57 Morehouse Lane
Red Hook, NY 12571

Some format issues inherent in the e-media version may also appear in this print version.

Printed by Curran Associates, Inc. (2019)

King Fahd University of Petroleum & Minerals
Dhahran 31261
Kingdom of Saudi Arabia

Phone: +966 (13) 860 0000

info@kfupm.edu.sa

Additional copies of this publication are available from:

Curran Associates, Inc.
57 Morehouse Lane
Red Hook, NY 12571 USA
Phone: 845-758-0400
Fax: 845-758-2633
Email: curran@proceedings.com
Web: www.proceedings.com

الندوة السنوية السعودية اليابانية الثامنة والعشرون

28th Annual Saudi-Japan Symposium

التقنية في تكرير البترول والبتروكيماويات

Technology in Petroleum Refining & Petrochemicals

PROCEEDINGS: 28

4-5 ربيع الأول 1440هـ

November 12-13, 2018

KFUPM Research Institute
King Fahd University of Petroleum & Minerals
Dhahran, Saudi Arabia

Preface

This is the Proceedings of the 28th annual symposium "*Technology in Petroleum Refining & Petrochemicals*" held on November 12-13, 2018 at King Fahd University of Petroleum & Minerals (KFUPM), Dhahran, Saudi Arabia. This event, jointly organized by KFUPM, Saudi Aramco, the Japan Petroleum Institute (JPI), and Japan Cooperation Center, Petroleum (JCCP), brings together researchers from industry and academia to discuss recent downstream technologies related to refining and petrochemicals.

At this year's symposium, there are 24 presentations featuring two keynote papers addressing hydrogen energy and future fuels/petrochemicals followed by 22 lectures on clean energy, catalyst technologies for aromatics, heavy oil and IMO sulfur regulations, numerical modeling, light olefins production, NMR utilization in R&D, and polymer applications. Speakers are from Saudi Aramco, KFUPM, SABIC, KAUST, JGC, JPI, Honeywell UOP, Japan Petroleum Energy Center and Japanese universities of Hiroshima, Hokkaido and Waseda.

We extend our thanks to KFUPM, Saudi Aramco, JPI, JCCP, distinguished speakers and attendees for their active participation and look forward to your participation in our next symposium.

Dr. Hassan S. Al-Asiri
Director, Center for Refining & Petrochemicals
http://ri.kfupm.edu.sa/crp/
KFUPM-Research Institute, Dhahran 31261, Saudi Arabia

Contents

PAPER 1 Keynote

Characteristics of hydrogen carriers and application to fuel cells

Koichi Eguchi
Department of Energy and Hydrocarbon Chemistry,
Graduate School of Engineering, Kyoto University, Kyoto, Japan
e-mail: eguchi@scl.kyoto-u.ac.jp Phone: +81-75-3832519

Hydrogen is considered a clean and efficient energy source for the near future. It is most desirable as energy systems that hydrogen is produced from renewable energy sources and consumed by direct combustion or by fuel cell systems. The importance of the energy carrier for the utilization of renewable energy has been recognized in recent years because long-term and stable storage of hydrogen should complete its supply chain in the form of energy carriers. Candidate substances for these carriers are methane, liquid hydrogen, organic chemical hydride, and ammonia. For transportation of long distance, storage of hydrogen or hydride in stable liquid form is required. Properties, production, transportation, and utilization depend strongly on candidate materials for energy carriers.

For conversion of energy carriers to electricity, combustion of hydrogen, methane, and ammonia has been investigated. Another method is to supply the carrier compound to fuel cells directly. In the case of ammonia and methane the efficiency can be enhanced when endothermic decomposition or reforming is coupled with high temperature exothermic electrochemical oxidation of solid oxide fuel cells (SOFCs). In the case of methane, coupling of reforming and SOFC has been known as internal reforming SOFC which has been adopted in many systems. Direct supply of ammonia to a fuel cell was possible leading to a simplified system with high-efficiency generation.

PAPER 2

Hydrogen production through Thermo-neutral reforming of liquid fuels using structured catalysts

Shakeel Ahmed[1], Sai P. Katikaneni[2], Uwais Baduruthamal[1], and Aadesh Harale[2]

[1] Center for Refining and Petrochemicals, Research Institute, King Fahd University of Petroleum & Minerals, Dhahran-31261, Saudi Arabia.

[2] Research & Development Center, Saudi Arabian Oil Company (Saudi Aramco), Dhahran, Saudi Arabia.

Hydrogen production is gaining increased importance from the viewpoints of ultimate clean and high calorific energy sources, and especially for the use in fuel cells. Hydrogen demand in the refining sector is also expected to continue its growth driven mainly by stricter fuel specifications. In addition, hydrogen-based fuel cells for automotive and stationary applications are gaining popularity for various reasons including their higher efficiencies and lower emissions. Use of liquid hydrocarbon fuels to generate hydrogen is being considered as an immediate solution for large scale hydrogen production. Liquid hydrocarbon fuels steam reforming is known for posing a major challenge due to its high carbon content, aromatics, and sulfur, which play a major role in catalyst deactivation. Similarly, the requirements for on-board hydrogen production demand a robust and highly efficient, compact, and cost effective reaction system for fuel processing. For this purpose, a multi-components thermo-neutral reforming (TNR) catalyst in various sizes and shapes was used to convert liquid fuels such as heavy naphtha and diesel into hydrogen rich syngas for fuel cell based power applications. Starting from powder form to pallets, monolith and finally microlith (brand name form PCI, USA) coated TNR catalyst was prepared, characterized and evaluated for syngas production. Monolith or microlith based structured catalysts are expected to withstand the rugged environment in which these fuel processers on-board vehicles are subjected for applications. In this study, structured TNR catalysts were prepared to achieve complete conversion of liquid fuels to hydrogen rich syngas.

Monolith Catalyst

Microlith Catalyst

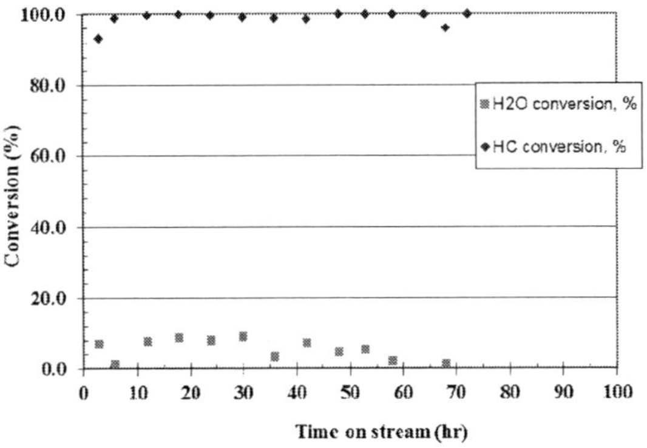

Performance evaluation of the Microlith-based TNR catalyst with Heavy Naphtha: Hydrocarbons and water conversion.

PAPER 3

Catalytic steam reforming of biomass-derivatives for hydrogen production

Yasushi Sekine
Waseda Univerity
Japan ysekine@waseda.jp

Catalytic steam reforming of bio-ethanol and bio-tar was investigated. As for bio-ethanol steam reforming, Co-catalyst showed higher performance than others. Loading of 1–2 wt% K on Co/α-Al2O3 was effective for improving catalytic activity, selectivity to hydrogen and suppressing byproduct (CH4, C2H4, and coke) formations. TEM measurements revealed that the Co0-CoO core–shell structure was formed over Co/K/α-Al2O3 during ethanol steam reforming at 823 K, suggesting that the oxidized Co species (CoO) is a highly active species in ethanol steam reforming. In-situ IR measurements revealed that the adsorbed ethanol forms stable acetate species by K loading, which improves hydrogen selectivity.

Applying electric field to the catalyst bed enables low temperature hydrogen production. Pt/CeO2 catalyst showed high activity in an electric field even at 423 K. We found that the supported platinum worked as an active site for the ethanol steam reforming. Conversion of ethanol and H2 yield drastically increased with imposing the electric field, and apparent activation energies for three elementary reactions (ethanol dehydrogenation, acetaldehyde decomposition, and acetaldehyde steam reforming) were lowered by the electric field.

For bio-tar steam reforming, Ni/perovskite catalysts showed excellent features including less coke formation. Ni/Al2O3 showed much coke formation, aggregated by oxidation or longer catalytic activity test, and lost its steam reforming activity by oxidation treatment. In contrast, Ni/La0.7Sr0.3AlO3-d catalyst showed high and stable steam reforming activity even after the oxidation treatment. Ni particles on Ni/La0.7Sr0.3AlO3-d retained the fine structure after oxidation treatment or reduction treatment.

PAPER 4

Design of scalable and stable catalysts for CO_2 reforming of methane

Oki Muraza
*[1]Center of Excellence in Nanotechnology and Chemical Engineering Department and, King Fahd University of Petroleum & Minerals, Saudi Arabia, *E-mail: omuraza@kfupm.edu.sa*

Natural gas has a crucial role in the energy transition, where low carbon resources are explored and produced in energy and chemical production. The successful production of shale gas in the North America has stimulated many countries to invest in unconventional gas. In addition to the industrial use of natural gas in ethane crackers to produce olefins and in the steam reforming to produce hydrogen, methane is also targeted due to its role as greenhouse gas (GHG). Together with carbon dioxide, another GHG, methane has been studied in different conversion routes such as methane-to-ethylene via oxidative coupling, methane-to-methanol via mild oxidation and CO_2 reforming of methane. The latter is an ideal process to convert two GHGs to synthetic gas (CO and hydrogen). Syngas has a strategic role as one of the seven feeds for petrochemical industries.

The CO_2 (dry) reforming is a potential route for natural gas valorization and mitigation of two greenhouse gases: CH_4 and CO_2. The most challenging task to have an industrial scale of dry reforming of CH4 is the severe coke deposition during reaction. We have evaluated operations on numerous catalysts with an emphasis potential catalyst supports to suppress coke formation or to design catalytic reactor with an automatic coke removal system. Other parameters in catalysis engineering such as reaction conditions, promoters, reactor design and novel regeneration system were highlighted. The bimetallic and tri-metallic catalysts were also reported elsewhere with respect to the role of metal-metal and metal-oxide support interaction. Base promoters such as potassium were also noted as the key to suppress coke formation. Perovskites as one of the most potential support was also noted, with its lost surface area limitation. In addition to their stability properties, scalability or the possibility to produce these catalysts are important for in commercial scale. In addition to silica, alumina, other metal oxides, zeolites, hydrotalcites, perovskites and pyrochlores were studied intensively as catalysts in CO_2 reforming of methane. Some of the most potential catalysts will be highlighted to shed a light in catalyst design for better CO_2 reforming catalysts.

References

1. A.R. Bawah, Z.O. Malaibari, O. Muraza, Syngas production from CO2 reforming of methane over Ni supported on hierarchical silicalite-1 fabricated by microwave-assisted hydrothermal synthesis, *International Journal Hydrogen Energy* 43 (2018) 13177-13189.
2. O. Muraza, A. Galadima, A Review of Coke Management in Dry Reforming of Methane, *International Journal Energy Research* 39 (2015) 1196-1216.

PAPER 5 Keynote

Future fuels & petrochemicals: new technologies in the face of new challenges

James Johnson
Honeywell UOP, USA

There are many trends that will provide challenges to the technologies of the industries that serve the transportation and chemicals markets. Some of these trends are more visible and dominant – such as increasing middle class populations and spending power, desire for mobility, aging population and changes in fuel demands in response to legislation. Other trends are more subtle but could still have great impact on these industries – for example plant security and reliability, environmental pressures on CO_2 footprints and emergence of EV's. The successful companies will be the ones who are well-positioned with technologies that anticipate and respond to these trends. This presentation explores these trends and shows how innovation and new capabilities can help refiners and petrochemical producers meet the upcoming challenges.

PAPER 6

Clean energy generation by artificial photosynthesis based on semiconductor technology

Kazuhiro Ohkawa
Professor of Computer, Electrical and Mathematical Sciences and Engineering Division,
Principal investigator of Energy Conversion Devices and Materials (ECO Devices) Lab.
King Abdullah University of Science and Technology (KAUST)
Thuwal 23955-6900, Saudi Arabia
E-mail: kazuhiro.ohkawa@kaust.edu.sa

Semiconductor technology is upgrading solar panels, light-emitting diodes (LEDs) and so on. We are developing optical devices by using nitride semiconductors. The material has a great potential to realize highly efficient LEDs and photocatalysts since its bandgap can cover the visible light region varying its alloy composition. We have invented a nitride photocatalyst that can generate solar hydrogen gas by water splitting, and hydrocarbons such as C_2H_5OH, HCOOH by artificial photosynthesis. This technology will be useful for clean energy generation and storage.

The author's first nitride photocatalyst patent (Pat. 3730142, the filing date: July 2001) became the world's first basic patent with nitride as a photocatalyst. [1]. The idea of the nitride photocatalyst came from the fabrication of lasers and LEDs with ZnSe or GaN. These materials did not have an appropriate etching solution, and Cl-based dry etching was used. In other words, these materials are chemically tough. Moreover, although their bandgaps (Eg) are large, the value of Eg becomes small in the case of ZnCdSe or InGaN alloy. Based on these characteristics, it was considered to be suitable for photocatalyst utilizing visible light. The electrochemical measurement of the band edge positions of GaN revealed that those are ideal as shown in Fig. 1 [2]. By using InGaN, the water reduction/oxidation (redox) levels exist between its conduction band and valence band edges. It suggests a possibility of high efficiency in photocatalysis by InGaN, and water splitting without extra bias.

Hydrogen generation by nitride photocatalyst was announced for the first time in 2005 in the world [2]. There were great responses in television and newspapers as having opened up new possibilities. However, there were problems with photo-corrosion and low efficiency. Since the efficiency is low, a voltage of 1.0 V was applied between the nitride working electrode and the metal counterelectrode. Now, the efficiency has been improved by controlling the depletion layer, and hydrogen can be generated even with zero applied voltage [3, 4].

Photo-corrosion was serious. The materials are chemically tough, but not enough for the case with light. Electron-hole pairs are created by photoabsorption. When electrons move to the counterelectrode side and hydrogen is generated, holes are left on the semiconductor side. When holes oxidize water, oxygen evolution occurs. However, if holes oxidize the semiconductor itself, it becomes a photo-corrosion [5]. At our first research stage, we chose hydrochloric acid, which can generate chlorine with higher priority than oxygen generation and photo-corrosion reaction, as an aqueous solution. In hydrochloric acid, it is hydrogen evolution but not water splitting.

Discovery of NiO cocatalyst against nitride photocatalysts has realized the higher efficiency of water splitting, and long durability in 2009 [6]. Figure 2 indicates that the

GaN-NiO photoelectrode shows better efficiency compared with GaN without NiO by about one order of magnitude. Durability was also improved, and the water splitting and artificial photosynthesis were realized [7, 8]. We confirmed durability for more than 500 hours (Fig. 3) [9]. We also demonstrated artificial photosynthesis of CO_2 reduction [10].

In the artificial photosynthesis, we observed the production of CO, HCOOH, CH_4, C_2H_4, C_2H_5OH, CH_3CHO, CH_2CHCH_2OH, C_3H_6OH, C_2H_5CHO from CO_2 reduction [11] (Fig. 4). Produced substances will be more because our current instruments limit the detection of molecules. The evidence of more materials is that the total Faraday efficiency of these substances is only about 80%. Moreover, we suspected whether it is derived from CO_2 molecule or not. We have experimented with the $^{13}CO_2$ gas consisting of carbon isotope with the atomic weight of 13 and confirmed its origin [11].

There are two major research subjects in future to achieve high efficiency. (1) Surface structure: for example, structures with a large surface area and high light absorption such as nanocolumns are desirable. The surface of nanocolumns contains different planes other than c-plane. Durability against these surfaces except for c-plane has not yet been secured. (2) Thick InGaN layer: the 100-nm thickness is necessary for sufficient photoabsorption in the InGaN layer, but the critical thickness of high-In-content InGaN is less than 100 nm. The thickness of 100 nm is same with the light penetration depth in nitrides, and the amount of light absorption becomes insufficient unless it has such thickness [4]. When we grow InGaN layers more than the critical thickness, electron-hole recombination becomes large due to the introduction of defects, and the efficiency of separating electrons and holes is significantly reduced. Currently, we can increase In content in InGaN layers on sapphire substrates until 6% [12-15]. If we might realize In contents of 20% - 30% with a thickness of 100 nm, efficiency will make a big leap. The emergence of InGaN substrates and/or the development of new epitaxy techniques such as InGaN/BGaN strain-compensation-structure [16] are crucial.

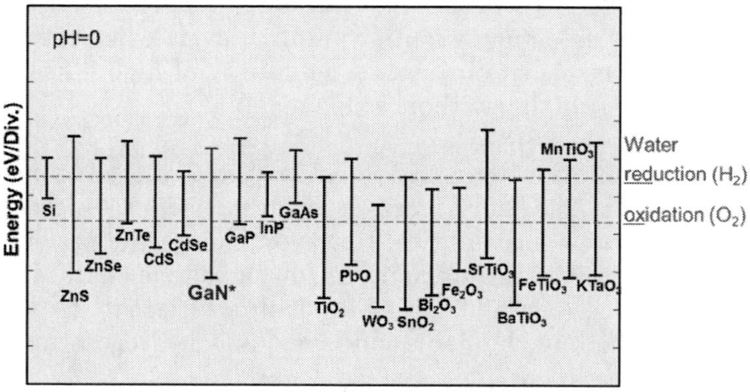

Fig. 1. The positions of the conduction band and the valence band position of various semiconductors. The water redox levels were included. The positions of GaN were originated from Ref. 2.

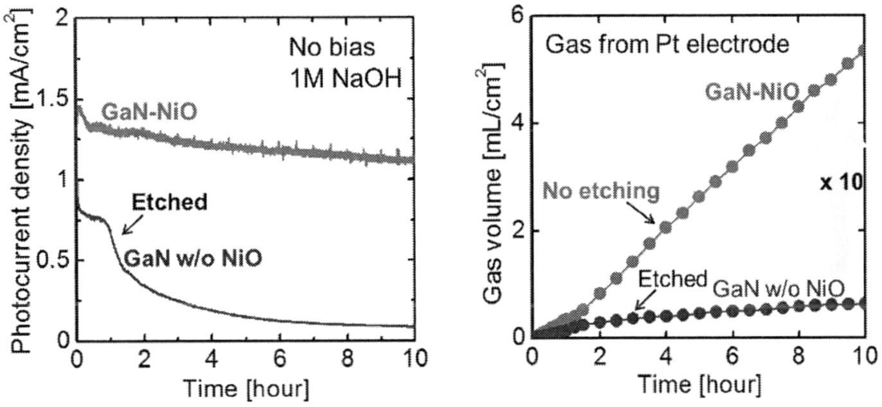

Fig. 2. GaN photocatalysis with and without NiO loading. The left shows photocurrent and the right shows hydrogen gas evolution. The GaN without NiO caused photo-corrosion from the initial stage and starts cutting off the current path in about 1 hour, so the photocurrent dramatically decreased. The light source was a Xe lamp, and the light intensity was 100 mW/cm².

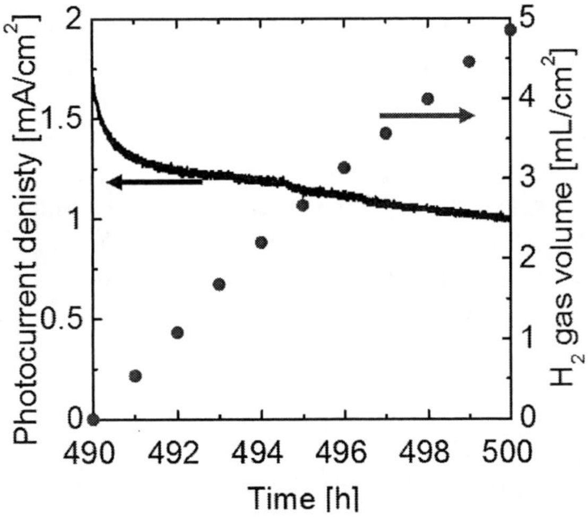

Fig. 3. Lifetime test of NiO-loaded GaN photocatalyst without bias [9]. The result is the 50th experiment of the 10-hour one. Stable H_2 generation was observed even after 500 hours. The sample structure was an n - GaN (n=1.2x10^{17} cm^{-3} at RT, 3 μm) layer grown on a c-plane sapphire substrate. The light source was a Xe lamp, and the light intensity was 100 mW/cm². The solution was 1 M NaOH aq.

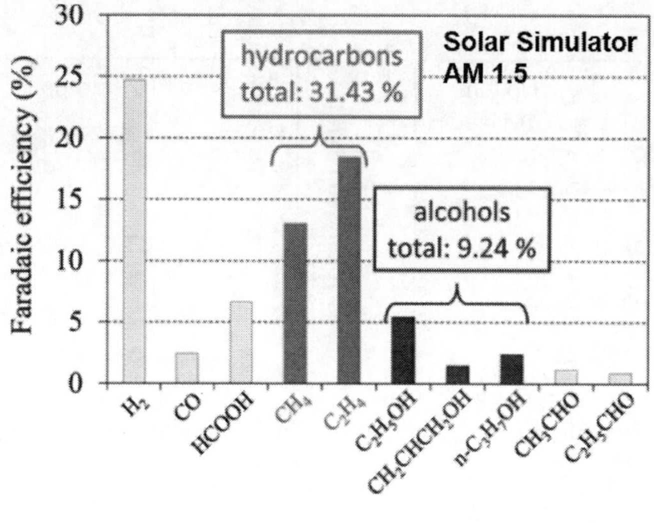

Fig. 4. An example of materials produced by artificial photosynthesis using nitride photocatalyst [11]. All of these products were from the reduction side. Solar-to-Fuels energy conversion efficiency was 0.408%. Faraday efficiencies to Hydrocarbons and alcohols were 31% and 9%, respectively. The light source was a solar simulator of AM 1.5.

References

[1] Kazuhiro Ohkawa, Japanese patent 3730142, the filing date: July 16, 2001.
[2] "Hydrogen gas generation by splitting aqueous water using n-type GaN photoelectrode with anodic oxidation", K. Fujii, T. Karasawa, K. Ohkawa, Japanese Journal of Applied Physics **44**, L543 (2005).
[3] "Photoelectrochemical reaction and H_2 generation at zero bias optimized by carrier concentration of n-type GaN", M. Ono, K. Fujii, T. Ito, Y. Iwaki, A. Hirako, T. Yao, K. Ohkawa, Journal of Chemical Physics **126**, 054708 (2007).
[4] "Nitride photocatalyst to generate hydrogen gas from water", Y. Iwaki, M. Ono, K. Yamaguchi, K. Kusakabe, K. Fujii, K. Ohkawa, physica status solidi (c) **5**, 2349 (2008).
[5] "Investigation of surface morphology of n-type GaN after photoelectrochemical reaction in various solutions for H_2 gas generation", K. Fujii, T. Ito, M. Ono, Y. Iwaki, T. Yao, K. Ohkawa, physica status solidi (c) **4**, 2650 (2007).
[6] 8th International Conference on Nitride Semiconductors, Oct. 18-23, 2009, Jeju, Korea, DD5. "H_2 Evolution from Water by GaN Photocatalyst with NiO Mictoelectrodes", F. Sano, T. Koyama, M. Sorimachi, A. Hirako, K. Ohkawa.
[7] "High stability and efficiency of GaN photocatalyst for hydrogen generation from water", T. Hayashi, K. Ohkawa, et al., Japanese Journal of Applied Physics **51**, 112601 (2012).
[8] "CO_2 conversion with light and water by GaN photoelectrode", S. Yotsuhashi, M. Deguchi, Y. Zenitani, R. Hinogami, H. Hashiba, Y. Yamada, K. Ohkawa, Japanese Journal of Applied Physics **51**, 02BP07 (2012).
[9] "Highly stable GaN photocatalyst for producing H_2 gas from water", K. Ohkawa, W. Ohara, D. Uchida, et al., Japanese Journal of Applied Physics 52, 08JH04 (2013).
[10] "Photo-induced CO_2 reduction with GaN electrode in Aqueous System", S. Yotsuhashi, K. Ohkawa, et al., Applied Physics Express **4**, 117101 (2011).

[11] "Analysis of products from photoelectrochemical reduction of $^{13}CO_2$ by GaN-Si based tandem photoelectrode", T. Sekimoto, H. Hashiba, S. Shinagawa, Y. Uetake, M. Deguchi, S. Yotsuhashi, K. Ohkawa, Journal of Physical Chemistry C **120**, 13970 (2016).

[12] "Photoelectrochemical properties of InGaN for H_2 generation from aqueous water", K. Fujii, K. Kusakabe, K. Ohkawa, Japanese Journal of Applied Physics **44**, 7433 (2005).

[13] "Tandem photo-electrode of InGaN with two Si p-n junctions for CO_2 conversion to HCOOH with the efficiency greater than biological photosynthesis", T. Sekimoto, K. Ohkawa,et al., Applied Physics Letters **106**, 073902 (2015).

[14] "InGaN photocatalysts on conductive Ga_2O_3 substrates", T. Ogita, Y. Uetake, Y. Yamashita, A. Kuramata, S. Yamakoshi, K. Ohkawa, physica status solidi (a) **212**, 1029 (2015).

[15] "Wireless InGaN-Si/Pt device for photo-electrochemical water splitting", T. Sekimoto, K. Ohkawa, et al., Japanese Journal of Applied Physics **55**, 088004 (2016).

[16] "Improvement of hydrogen evolution from InGaN/BGaN photoelectrodes", K. Shiohama, W. Ohara, H. Hayashi, D. Uchida, K. Ohkawa, Annual meeting of Japanese Society of Applied Physics, Sep. 16-20, 2013, Doshisya Univ., Kyoto, 16p-B5-3.

PAPER 7

The effect of lanthanum exchange on USY-zeolites for the alkylation of 2-butene with isobutane

Mohammed R. Al-Dossary
Research & Development Center
Saudi Aramco, Dhahran

The growing demand for higher quality gasoline coupled with stringent environmental regulations requires the reduction in aromatics, sulfur and nitrogen compounds. The alkylation process utilizes low value olefins with isobutane producing high octane gasoline that improves the refinery's profit margin.

Current technologies utilize harmful homogenous catalysts (hydrofluoric or sulfuric acids) as catalysts for the reaction. Those acids present serious health and environmental concerns in addition with the high operational and mitigation costs. Zeolites present an ideal substitute to both acids due to the benign nature and acidic properties of zeolites. Faujasites in particular are very attractive catalysts for the alkylation reaction due to their large-pores and high acidity which can be tailored towards alkylation. Deactivation remains a major challenge for a zeolite-based alkylation process.

The effect of lanthanum (La) exchange on the activity, stability and selectivity of the faujasite USY was studied using a well-mixed, continuous Berty-type reactor at 75°C, 17 bar, an olefin space velocity of 0.30 h^{-1} and an isobutane/2-butene ratio of 15 (mol/mol). The exchange significantly improved the catalytic stability with the La-exchanged zeolites reaching higher butene conversions by 25 % after 10 hours on-stream mainly due to the reduction in strong Lewis acidity. The optimum number of lanthanum exchanges was three exchanges followed by calcination creating the highest ratio of strong-Brønsted/strong-Lewis ratio.

PAPER 8

Growing role of residue upgrading process
- Responding to the IMO 2020 SOx regulation –

Atsunori Sato
Technology Innovation Center, Process Technology Division,
JGC Corporation, 2-3-1, Minato Mirai, Nishi-ku, Yokohama 220-6001, JAPAN

1. Abstract

According to the decision of the International Maritime Organization (IMO), sulfur oxides (SOx) emissions regulation from ships in general waters will become strict from January 2020. To comply with the IMO SOx regulation, the shipping company should convert existing high sulfur residual fuel oil (HSFO) containing a maximum of 3.5wt% sulfur into low sulfur fuels, typically fuel oil or LNG, having a sulfur content of less than 0.5wt%, or it is necessary to install a device (SOx Scrubber) to remove SOx from exhaust gas if HSFO is still used. Most oceangoing ships should shift to the low sulfur fuel oil since the number of SOx Scrubbers installed by 2020 will be very limited due to physical restrictions. As a result, the demand for HSFO for marine fuel will decrease sharply after enforcement of the IMO SOx regulation.

Demand for HSFO for marine fuel is about 4.5 million barrel/day, accounting for about 5% of the global demand for crude oil. Refineries responsible for fuel supply are under pressure to reduce HSFO production and supply the low sulfur fuel oil. Vacuum residue (VR) is the main base material of HSFO, and the importance of residue upgrading processes will be further enhanced in order to reduce and convert VR to the lighter petroleum products which are in demand.

Considering these circumstances surrounding the IMO SOx regulation, in the first half of this report, the general measures in response to the IMO SOx regulation and the current directions of responses to the regulation by shipping companies and refineries are introduced. In the latter half of this report, economics of major residue upgrading processes are compared by considering future petroleum products market and some of the latest trends of residue upgrading technologies are introduced.

2. Measures for the IMO SOx Regulation

2.1 Technical Options and Issues

The three main measures by ship operators to the IMO SOx Regulation and their respective issues are shown in Figure 1. The first method is to substitute the existing HSFO with the low sulfur fuel oil. The second method is to install SOx Scrubber in the exhaust gas line of the ship and continue to use HSFO. The third one is to serve LNG fuel ship. Although it is technically possible to deal with any case, there are some issues in each case. In the case of substituting with the low sulfur fuel oil, it is necessary to secure a sufficient quantity of the fuel oil and to standardize the quality. Regarding fuel quality, standardization by ISO is under way. In SOx Scrubber installation, additional investment for ship modification is necessary, and the number of ships that can have them installed by 2020 is very limited. With respect to LNG fuel carriers, it takes time to spread it since LNG fueled vessels can be applied mainly to new shipbuilding, the cost of such ships is higher than for those using fuel oil and an infrastructure for suppling LNG fuel for ships is not well -established worldwide.

Measures	Switch to Low Sulfur Fuel Oil (< 0.5wt% Sulfur) (Use Existing Engine)	Use HSFO with SOx Scrubber (Currently≒600 Ships)	LNG Fuel Ship (Currently≒100 Ships)
		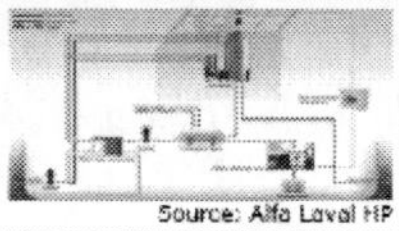	
Issues	·Secure supply ·Quality standardization ·Uncertain fuel price	·Limited dock number ·Secure installation space ·Additional investment cost (a few million US$)	·Lack of fuel supply infrastructure ·New shipbuilding only ·Higher shipbuilding cost

Figure 1 Three Measures by Ship Operators and Issues to the IMO SOx Regulation

2.2 Marine Fuels Market

2.2.1 Trend in Response Measures

Application of LNG fueled ships is basically limited to new shipbuilding, and the remaining two options can also be applied to a major portion of existing ships. The marine fuel market will change depending upon which measure the existing ships take. Table 1 summarizes the estimated number of existing ships in 2016 subject to the IMO SOx regulation based on our interviews with Japanese shipping industry. According to the IMO survey[1)], installation of SOx Scrubbers is estimated to be feasible in about 90% of new ships and about 60% of existing ships. In the Table, the number of ships economically viable for SOx Scrubber installation is estimated conservatively by taking 60% of all the regulated ships, i.e. 55,000 ships. According to the survery[2)] by the Japan Petroleum Energy Center (JPEC), the number of ships with SOx Scrubbers by 2020 is estimated to be 1,500 to 2,500 ships as maximum, which is the same range with our survey conducted for Japanese shipping companies.

Table 1 Number of Ships and Potential SOx Scrubbers Installation

Ships	Estimated Number (2016)
Ships subject the IMO SOx Regulation (Among them, ships passing through the Emission Control Area, ECA)	55,000 (10,000)
Ocean going ships	40,000
Domestic ships	15,000
Potential Ships of economically viable for SOx Scrubber installation	33,000 60% of 55,000 ships
Ships with SOx Scrubber by 2020	1,500~2,500

If SOx Scrubbers are introduced to a major portion of all the regulated ships by 2020, the major portion of the ships will be able to use the current HSFO. In this case, the impact on the marine fuel market will be moderated. However, the proportion of ships with SOx Scrubbers is small when compared to the total number of regulated ships, so the major portion of the ships will have to take the low sulfur fuel oil option. In the latter case, the impact on the marine fuel market will be great.

2.2.2 Forecast of Marine Fuels Market

Figure 2 shows an example of the demand forecast for marine fuels. There are some forecasts for the marine fuels market by multiple research organizations and almost the same predictions are made excepting how much the demand of HSFO returns after 2020. Most ships without SOx Scrubbers in 2020 will need to take the low sulfur fuel oil, substituting for HSFO. As a result, the demand for HSFO will decline rapidly in 2020. After 2020, the introduction of SOx Scrubbers gradually progresses, and the demand for HSFO will recover. It is difficult for refineries to make concrete countermeasures for the marine fuels supply since the market is unclear in terms of the recovery of HSFO, and the demand for, and price of the low sulfur fuel oil such as diesel/Gas Oil, ultra-low sulfur fuel oil (ULSFO) containing less than 0.1wt% sulfur and low sulfur residual fuel oil (LSFO).

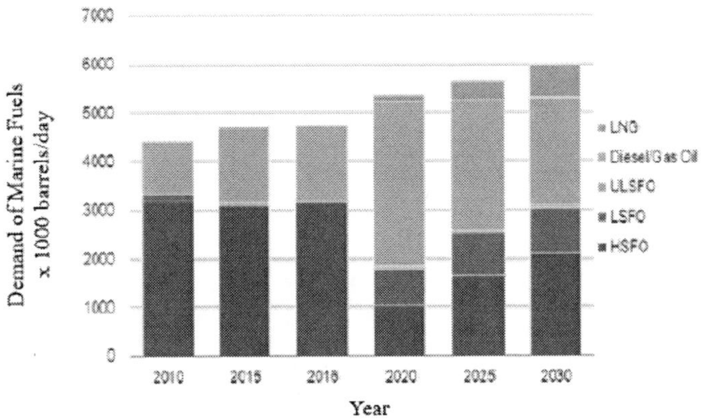

Source: Nexant, 2017

Figure 2 Forecast of Marine Fuels Demand[2)]

2.3 Trend of Measures by Refineries

2.3.1 Refineries in Japan

In response to the IMO SOx regulation, Cosmo Energy Holdings[3)] has announced revamping of delayed coker aiming at reduction of residual oil production in its Sakai Refinery. There is no other plan that is specifically disclosed by Japanese refineries. In Japan, demand for petroleum products is decreasing with the declining population, so each refinery would not make a huge investment in Japan. Under such circumstances, the following options would be considered toward reduction of residue and/or conversion of residue to demanding lighter fractions.

- Revamping of residue hydrodesulfurization process (RHDS)
- Revamping of residue conversion process (for example, delayed coker expansion by Cosmo Oil)
- Installation of Solvent De-asphalting (SDA) process to reduce VR

2.3.2 Refineries in Europe, the United States and Asia

(1) Europe

Based on a survey[2)], the trend of measures by the refineries in Europe, the United States and Asia are introduced as follows. It can be said that the measures by European refineries are considered to be most progressed. For example, the following measures are taken:

- Maersk Line (Denmark, the world largest container logistics company) and CMA CGM (France, the world third largest container logistics company) announced the adoption of distillate oil. CMA CGM has signed a fuel supply contract with Total.
- ExxonMobil announced guidelines stating handling procedures for low viscosity fuel product in 2018. It is expected to market the low sulfur fuel oil by the beginning of 2019.
- CEPSA also plans to sell the low sulfur fuel oil from the third quarter in 2019.
- As shown in Table 2, several refiners in Europe are considering capital investment for residue upgrading processes. In particular, there are many studies regarding the introduction of Solvent De-asphalting (SDA) and Hydrocracker (HDC).

Table 2 Investment Plans in Residue Upgrading by European Refineries[2)]

Country	Company	Refinery	Process	Capa(kBD)	Completion
Finland	Neste	Porvoo	Solvent Deasphalting(SDA)	10	2017 2Q
Belgium	Total	Antwerp	Solvent Deasphalting(SDA)	48	2017 3Q
Belgium	Total	Antwerp	Mild Hydrocracker	20	2017 3Q
Norway	Exxon Mobil	Slagen	VDU/Residue flash tower	35	2017 3Q
Belgium	Exxon Mobil	Antwerp	Coker	50	2017 4Q
Spain	BP	Castellon	VDU/revamp	14	2018 1Q
Poland	Grupa Lotos	Gdansk	VDU	21	2018 1Q
Poland	Grupa Lotos	Gdansk	Coker	26	2018 1Q
Holland	Exxon Mobil	Rotterdam	Hydrocracker replacement	15	2018 4Q
Sweden	Preem	Lysekil	VDU	36	2018 4Q
Holland	Shell	Pernis	Solvent Deasphalting(SDA)	30	2018 4Q
Serbia	Gazprom/NIS	Pancevo	Coker	13	2018 4Q
France	Total	Donges, Loire	VGO hydrotreater	40	2019 4Q
Croatia	INA/MOL	Rijeka	Coker	15	2019 4Q
Poland	PKN Orlen	Plock	RHDC	12	2019 4Q
Poland	PKN Orlen	Plock	Visbreaker	20	2019 4Q
Holland	Total/Lukoil	Vlissingen, Flushing	HDC	18	2020 2Q
Germany	Shell	Wesseling(Rheinland)	Solvent Deasphalting(SDA)	14	2020 3Q
Spain	Cepsa	Gibralter San Roque	RHDC	43	2021 2Q
Sweden	Preem	Lysekil	RHDC	42	2022 2Q
Lithuania	PKN Orlen	Mazeikiu	RHDC	40	2022 3Q

Source: Made by JPEC based on various information

(2) The United States

According to the report[4)] in 2018 by the U.S. Energy Information Administration (EIA), the refinery capacity in Petroleum Administration for Defense Districts 3 (PADD3) is 10.43 million barrels per stream day (BPSD). The refinery capacity in PADD3 shares 53% of the national refining capacity and 56 refiners are in operation in the PADD.

The refineries in PADD3 are basically equipped with residue upgrading processes such as cokers and it is considered that they could supply the low sulfur fuel oil in 2020. Main measures taken by the refineries in PADD3 are divided into the following four categories. The refining companies belonging to each category are shown in Table 3[2)].

Category 1: It is producing ULSFO or another marine fuel oil.

Category 2: It is producing ULSFO and another low sulfur fuel oil.

Category 3: It is taking measures for multiple types of marine fuels supply. Supply of LNG marine fuel has been announced.

Category 4: It is not producing marine fuel oil and there are no production plans for the future.

Table 3 Measures by the Refineries in PADD3[2)]

Category	Company Name
1	Andeavor*, Calumet Specialty Products, Deer Park Refining, Delek U.S., Ergon Inc, Hunt Consolidated, LyondellBasell, Marathon Petroleum Corp*, Motiva Enterprises, PBF Energy Co, PDV America, Petrobras America, Phillips 66, Total S.A., Transworld Oil U.S.A., Valero Energy Corp
2	Chevron, Exxon Mobil, Goodway Refining, Martin Resource Management Group, Royal Dutch Shell
3	Exxon Mobil, Royal Dutch Shell, Total S.A.
4	Blue Dolphin Energy Co, Buckeye Partners, Excel Paralubes, Hollyfrontier Corp, Kinder Morgan Energy Partners, Koch Industries, Magellan Midstream Partners, Petromax Refining Co, Placid Oil Co, Texas Oil & Chemical Co, WRB Refining

Source: International Technology and Trade Associates, Inc., Washington, DC

(3) Asia (China, Korea and Singapore)

There is little specific information published on the refineries measures of Singapore, Korea and China. Based on the information of supply and demand forecast of marine fuels[2)] as shown in Figure 3, it is considered that the following measures would be taken in each country:

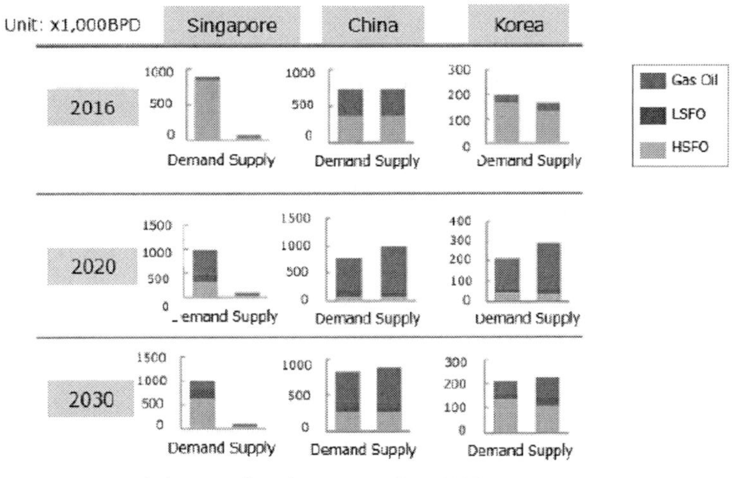

Source: Made by JPEC based on Nexant data, 2017

Figure 3 Supply and Demand of Marine Fuels in China, Korea and Singapore[2)]

In Singapore, most marine fuels are depending on imports since domestic supply is limited. China is considered to increase the production of diesel / gasoil for marine fuel as the demand for the fuels will increase in 2020. The trend in Korea is similar to that of China. As SOx Scrubbers continue to spread toward 2030, the demand for HSFO will recover and the supply of diesel / gasoil will decrease.

3. Refining Technologies to the Future Petroleum Products Market

3.1 Future Petroleum Products Market

Figure 4 shows a forecast of petroleum products demand in the world[2)]. Demand for petroleum products is projected to grow at an annual rate of about 1% until 2030. Among the products, demand growths of LPG and naphtha used for petrochemical feedstocks are expected to be strong. Gasoline and middle distillates such as jet fuel and diesel will also increase steadily. However, growth in gasoline will slow down after 2020.

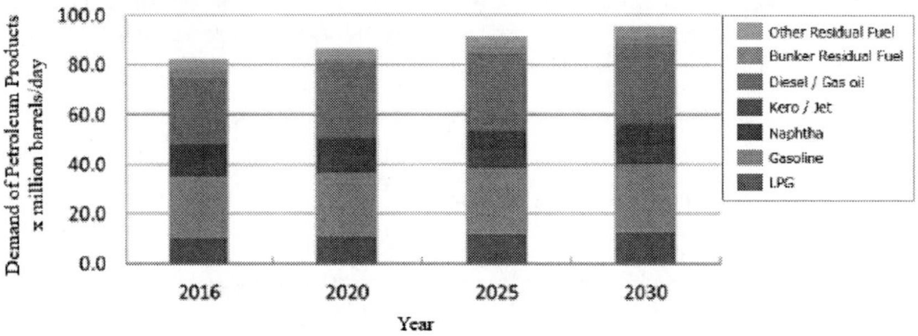

Source: Made by JPEC based on Nexant data, 2017

Figure 4 Forecast of Global Petroleum Products Demand[2)]

The increase in demand for residual fuel is small compared with the other petroleum products. The need for conversion from residue to high demand products such as middle distillates and naphtha is increased, and the need for residue upgrading processes is expected to be further enhanced.

3.2 Trend of Residue Upgrading Processes

3.2.1 Refineries Options to the IMO SOx Regulation

Table 4 shows the main options and potential measures by refineries to the IMO SOx regulation. Regarding the first option, it is considered that replacement of crude oil feedstock by that containing lower sulfur would be a costly option. Concerning the second option, a large amount of investment to reduce sulfur in HSFO is economically uncertain since the demand for HSFO is predicted to recover. Therefore, it would be most likely to consider the residue upgrading processes to convert residue to the lighter petroleum products which are in demand.

Table 4 Refineries Main Options to the IMO SOx Regulation

No.	Options	Potential Measures
1	No capital investment and continue business as usual	Change to lower sulfur crude oil to reduce sulfur in fuel oil (costly option)
2	Invest to reduce sulfur in marine fuel oil	Invest in residue hydrodesulfurization
3	Exit marine fuel oil market	Invest in residue upgrading process to convert demanding petroleum products

3.2.2 Economics of Bottom Upgrading Processes

The economics of residue upgrading processes have been evaluated by Chevron Lummus Global (CLG), one of the major licensors familiar with hydroprocessing and residue upgrading processes, and the study results[5)] were introduced at the Petroleum Refining Conference of the Japan Petroleum Institute (JPI). In the study, the economics of major residue upgrading processes were compared to the base configuration with VR processed by visbreaker. The crude oil feedstock was a mixture of half each of Arabian Light and Arabian Heavy. The study result is simplified and summarized in Table 5. It is evaluated that the value creation is higher if the ratio of the increment in net present value (Delta NPV) to the total investment cost (TIC) is higher. The study shows that value creation is the highest in the case of slurry phase hydrocracker (Slurry-HDC) following the case of ebullated bed hydrocracker (EB-HDC). LC-MAX® is the process incorporating SDA into LC-FINING ® (CLG's EB-HDC) and is introduced in more detail below. LC-SLURRY is CLG's Slurry -HDC.

Table 5 Economics Comparison of Residue Upgrading Processes

Case	Vacuum Residue Upgrading Process	Relative Total Investment Cost (TIC)	Delta NPV/TIC
Base	Visbreaker	-	-
Coker	Delayed Coker	$$$$$$$$	0.42
RFCC	RHDS + RFCC	$$$$$$$$$$$	0.42
LC-MAX®	LC-MAX® (EB-HDC +SDA)	$$$$$$$$$$$$	0.51 (higher)
LC-SLURRY	LC-SLURRY (Slurry-HDC)	$$$$$$$$$$$	0.58 (highest)

Source: Made by JGC based on CLG data, 2016[5)]

The periods of investment recovery in all cases are not much different, i.e. between 4 and 5 years. Although the investment costs are higher in the hydrocracker cases, the value creation is higher than the others due to higher yield of the lighter fractions. In the case of residue fluidized catalytic cracking (RFCC), the value creation is lower than the hydrocracker cases due to the catalyst cost which depends mainly on metals contents in the feed. As demand forecasts show that middle distillates, naphtha and LPG are expected to grow more greatly and hydrocrackers can produce those products for which there is demand with the highest yield, the following section introduces recent topics of residue upgrading processes mainly focusing on hydrocracking processes.

3.2.3 Recent Topics of Residue Upgrading Processes

(1) Ebullated Bed Hydrocracking Process

EB-HDC has been spreading as a useful technology for converting VR to lighter fractions mainly composed of middle distillates. LC-FINING® by CLG and H-OIL®$_{RC}$ by Axens are the major processes of EB-HDC. Due to the instability of the unconverted oil (UCO) produced in the process as residual oil, the conversion of VR is normally controlled in the range of 60% to 70%. The processes have been developed further to achieve higher VR conversion by incorporating SDA into EB-HDCs to remove asphaltene that is a cause substance in the formation of sediment. The main features of the technologies by CLG and Axens are described below.

Simplified process scheme[6] of LC-MAX® by CLG is shown in Figure 5. The process has two-stage reactors. The conversion of the first stage reactor is mild to be around 60% or less, and ashphaltene is removed from the effluent from the first stage reactor by SDA. The de-asphalted oil (DAO) is sent to the second stage reactor and processed further at a high conversion rate of 80% to 90%. 85% to 90% of VR conversion is achieved in the whole process.

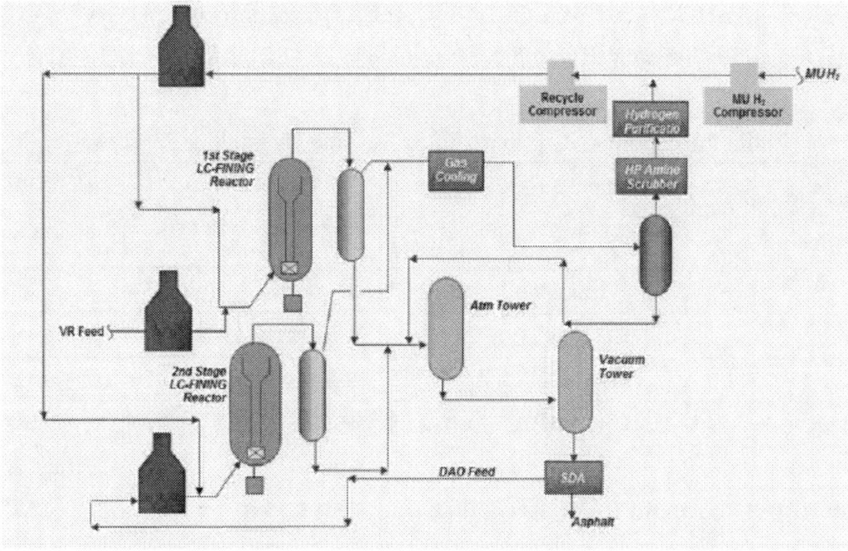

Source: Reproduced with permission from CLG

Figure 5 LC-MAX® Scheme (CLG)[6]

Axens has a scheme[7] that achieves a high VR conversion of 90% by incorporating Solvahl™ (Axens's SDA) into H-Oil®$_{RC}$. In the scheme, Solvahl™ removes unstable asphaltene from the reactor effluent and SDA pitch is gasified.

(2) Slurry Phase Hydrocracking Process

Table 6 shows the main characteristics and commercialization status of the Slurry-HDCs which have been publicly announced for commercialization activity such as commercial operation, construction and the conclusion of license agreements. At the present time, only EST has experience of commercial operation in ENI's refinery. The number of conclusions of the license agreement has been increasing, and recently Sinopec concluded a license agreement with ENI EST. Each company has been continuously pursuing developments for catalyst cost reduction and effective utilization of pitch generated a few percentages from the process. In Uniflex™, UOP has been studying the utilization of the pitch as cement raw material and boiler fuel.

Table 6 Features of Slurry Phase-HDCs and Commercialization Status

Process	EST (1)	VCC TM (2)	HDHplus®	Uniflex TM
Licensor	Eni	KBR, BP	PDVSA INTEVEP	UOP
Feed	VR	VR, Coal Tar	VR	VR
Catalyst	MoS_2	Fe /Fine Carbon	Mo & Ni Emulsion /Fine Coke	$FeSO_4$ /Fine Powder Coal
Conversion, %	92~94 (VR/Cat recycle)	95 (no recycle)	82~95 (97 by recycle)	90~95 (no recycle)
Pressure, Mpa	15~16	18~23	19~21	14
Temperature, ℃	425~435	450~	430~460	435~470
Commercial Operation, BPSD	Eni Sannazzaro at 23,000			
Construction, BPSD		- Tatarstan TAIF at 55,000 of VR	PDVSA Puerto La Cruz at 25,000 x 2 trains	
License, BPSD	- Total - Sinopec Zheijiang at 60,000 x 2 trains	- Jordan JPRC at VR - Others in China at 10,000 x 2 projects		- Pakistan NRL at 50,000 - Discussion in China, Russia

(1) Eni Slurry Technology (2) VEBA Combi-Cracking

(3) Utilization of SDA

SDA has a key role in recent refineries, as the process can be used in a variety of ways for residue upgrading. It is less expensive to build and operate than other residue conversion processes and is especially useful in recovering large quantities of high-quality oil (de-asphalted oil, DAO) from heavy hydrocarbon streams such as VR. In addition to the application in EB-HDC as mentioned above, SDA can remove useful DAO from VR. DAO can be a feedstock for the production of gasoline and/or middle distillates. Figure 6 shows a typical residue upgrading scheme with residue hydroprocessing (RHDS) plus RFCC. By introducing SDA as shown in Figure 7, DAO can be converted to value added lighter petroleum products, mainly to gasoline. It is one of the corresponding measures with minimum modification cost in the existing refineries where residual oil needs to be reduced and gasoline demand is declining.

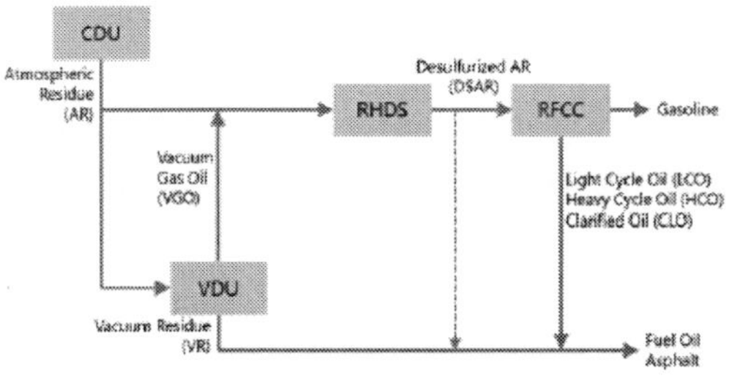

Figure 6 RHDS-RFCC Scheme

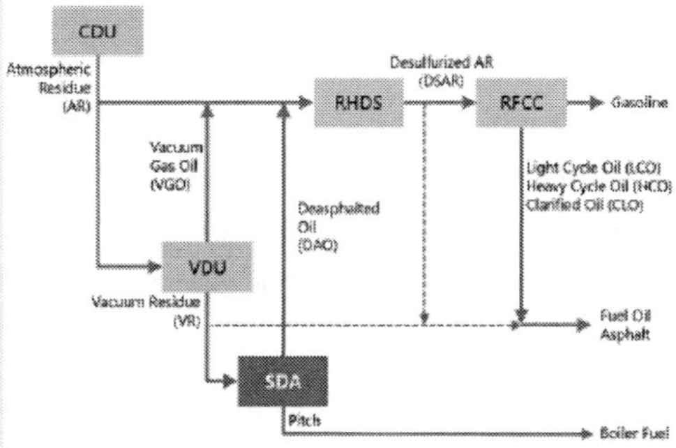

Figure 7 SDA Addition to RHDS-RFCC Scheme

In SDA, the softening point of the pitch is high and expensive cutter stock is used to lower the softening point. Quadrise has a technology[8)] to produce a stable orimulsion fuel named Multiphase Superfine Atomized Residue (MSAR) and has commercial experience for a boiler fuel. MSAR production plant is modularized, compact, occupies a small site area and has a short delivery time.

Kawasaki Heavy Industries[9] in Japan has been developed Kawasaki Advanced Clean Combustion Boiler (Upgraded KACC Boiler) which can use SDA pitch as a fuel oil without cutter stock.

4. Conclusion

- Shipping companies' responses toward the IMO 2020 SOx regulation in general waters are not simple. Refineries responses to the IMO SOx regulation are also not clear since the future demands and prices of marine fuels cannot be predicted accurately.
- Although it is such an uncertain situation, most ships will need to convert their marine fuel from HSFO to the low sulfur fuel oil after 2020. Refineries will need to reduce HSFO and supply more the low sulfur fuel oil.
- In response to the future petroleum products market, Japanese refineries are reviewing revamping of residue upgrading processes. Refineries in Europe have investment plans in new residue upgrading processes. The role of residue upgrading processes is growing.
- While the prediction of the petroleum products market is complicated, JGC can provide selected life cycle assessment of refineries by taking advantage of our experience in the construction of 5.25 million BPSD of CDU, equivalent to about 6% of world refining capacity.

5. References

1) CE Delft, STRATAS ADVISORS, UMAS, NMRI, PETROMARKET, 2016. "Assessment of Fuel Oil Availability Final Report". Website of CE Delft Available at https://www.cedelft.eu/publicatie/assessment_of_fuel_oil_availability/1858

2) JPEC, Japan Petroleum Energy Center, 2018. "海外における IMO 燃料油品質規制対応の動向調査". JPEC NEWS 2018 年 5 月. Website of JPEC Available at http://www.pecj.or.jp/japanese/jpecnews/pdf/jpecnews201805.pdf

3) Cosmo Energy Holdings HP. Website of Cosmo Energy Holdings Available at https://ceh.cosmo-oil.co.jp/

4)EIA, The U.S. Energy Information Administration, 2018. "Number and Capacity of Petroleum Refineries". Website of EIA Available at https://www.eia.gov/dnav/pet/pet_pnp_cap1_dcu_nus_a.htm

5) CLG, Chevron Lummus Global, 2016. "Innovative technology Solutions to Impending International Maritime Organization Regulations". JPT Petroleum Refining Conference

6) Chevron USA, 2014. "Recent Advances in Residuum Processing". JPI Petroleum Refining Conference

7) Axens, 2012. "Conversion Options for Clean Fuel Production". Petrofed-Petroleum Refining Technologies Seminar

8) Quadrise, 2016. "MSAR Refinery Technology". 14th International Bottom of the Barrel Technology Conference

9) Kawasaki Heavy Industries. Website of Kawasaki Heavy Industries Available at http://global.kawasaki.com/en/

PAPER 9

Effect of zeolite Y modification on reaction performance of heavy crude oil hydrocracking

Lianhui Ding
Research & Development Center, Saudi Aramco

Hydrocracking is one of the most important heavy oil conversion processes in modern refineries and petrochemicals plants. Hydrocracking catalysts play a critical role in determining the product slates and properties, and thus the economics. Zeolite Y is widely used cracking component of the commercialized hydrocracking catalysts. Its pore structure and acidity can be modified by various chemical or steaming treatments, and have critical effect on the catalyst reaction performances. In present study, zeolite Y was modified by different methods to generate modified zeolites with varied silica-to-alumina ratios, pore volumes, pore sizes, relative surface areas, and acidity. The reaction performances in hydrotreated crude oil were evaluated with these modified zeolite based NiMo catalysts in a high through-put reactor system. The results showed that, to achieve the optimum performance in heavy oil conversion, the pore structure and acidity of the catalysts need be well balanced.

PAPER 10

Oil soluble dispersed catalysts for slurry phase hydrocracking of heavy VGO

Tareq A. Al-Attas[1], Emad A.S. Bdwi[1], Syed A. Ali[2], Saad A. Al-Bogami[3], Mohammad M. Hossain[1]*

[1]Department of Chemical Engineering, [2]Center of Refining & Petrochemicals
King Fahd University of Petroleum and Minerals, Dhahran 31261, Saudi Arabia
[3]Research & Development Center, Saudi Aramco Oil Company, Dhahran, Saudi Arabia

Abstract

Present study deals with the promotional effects of dispersed co-catalysts on hydrocracking of vacuum gas oil (VGO). In this regard, oil soluble Mo, Fe and Co based materials are investigated. The experiments are conducted in a batch autoclave reactor (at 4 MPa and 420 °C) with and without the presence of a commercial first-stage hydrocracking W-Ni/Al_2O_3-SiO_2 catalyst. It is observed that the dispersed metal catalysts enhance hydrogenation activity and minimize the amount of coke formation. The lowest coke formation is achieved by using only 500 ppm dispersed catalysts in the feed. Among the three catalysts, Co and Mo based catalysts show lower amount of coke formation than that of Fe catalyst. It is observed that the addition of only 500 ppm of Co/Mo decreased the coke from 2.5 wt % to 0.9 wt % during the thermal run. The addition of dispersed catalyst together with the solid catalyst also shows similar decrease of coke formation. The presence of dispersed catalysts also enhances the percentage of naphtha and decreased the heavy VGO fraction when supported metal used. Based on the product analysis, a kinetic model was developed based on the experimental data for the catalytic test (dispersed + solid). The activation energy of VGO cracking forming distillate requires the least activation energy (1.477 kcal/mol) as compared to the other competing reactions, so VGO will most likely crack to distillate then distillate will crack to naphtha and naphtha will be cracked to gases.

PAPER 11

Development of *Petroleomics* from Fundamentals to Applications

Kazuhiro Inamura
Technology & Planning Department, Japan Petroleum Energy Center (JPEC)
E-mail: ka-inamura@pecj.or.jp

ABSTRACT

We, JPEC and our partners, have been challenging to develop "Petroleomics" as a new refining technology since fiscal year (FY) 2011, where "Petroleomics" can give us the way of ultimate approach based on molecular reaction models with molecular level analyses on heavy oil. After five years endeavor in the fundamental stage, our petroleum informatics database covered up to 25 million of chemical structures of heavy oil components had been constructed with the aid of ultrahigh resolution mass spectrometer (FT-ICR MS), and subsequently reaction modeling studies were applied Residue Hydrodesulfurization (RDS) and the aggregation model for asphaltenes to sediments, both of which are particularly important subjects of heavy oil upgrading.

Now, our Petroleomics project is in the application stage and tackling with three major tasks: molecular database of crudes including unconventional oils, total optimization of RDS and Residue Fluid Catalytic Cracking (RFCC) operations with reaction modeling, and mechanism of asphaltene aggregation responsible for fouling and plugging in some heavy oil upgrading processes. Further progress of Petroleomics is expected to achieve its practical usages in refineries, such as much better performance diagnosis, operational optimization and catalyst & process development.

1. Why we are challenging the development of *Petroleomics* as a new refining technology

Even in the field of refining technology, progress has been made by advances in modern analytical methods. Conventional Gas Chromatography (GC) can provide us both of qualitative and quantitative information of almost all chemical species in gasoline (Naphtha). GC methods have been made a lot of progress so that the most refractory sulfur compound in Gas oil (diesel) was identified as 4,6-Dimethyldibenzothiophene by a selective sulfur detector, Sulfur Chemiluminescence Detector (SCD), on GC in the early 2000s, and later two-dimensional gas chromatography (2D-GC or GC×GC) provided detail chemical species distributions in molecular level resolution from lighter paraffines to complex naphtenic tri-aromatics [1]. These analytical techniques have been contributed to achievements of both regulations and their refining processes for clean fuels, like lower benzene content in gasoline and ultra-low-sulfur gasoline and diesel [2]. However, heavy oils, like vacuum gas oil (VGO) and vacuum residue (VR), had not been identified at molecular level until quite recently, mainly because of their complexities. Fourier transform ion cyclotron resonance mass spectrometer (FT-ICR MS) as ultrahigh resolution MS was first applied as a tool for petroleum analysis by Dr. Hsu and coworkers at Exxon in 1994 [3], and later it has been greatly practiced by a research group at the National High Magnetic Field Laboratory (NHMFL) for heavy oils, those who called their methodology for petroleum as "*Petroleomics*" [4, 5, 6].

In Japan, as almost all crudes were imported, especially 87.3% from Middle East (39.4% from Saudi Arabia and 24.8% from United Arab Emirates) in 2017, efficient utilization of petroleum has been the most crucial issue in the oil industry. Among all, noble use of heavy oils, in other word, the challenge of technological enhancement of heavy oil upgrading has been the top target especially at JPEC. However, the conventional approach for the process studies, for example Residue Hydrodesulfurization (RDS) unit, considered heavy oils as simplified lumping models and overall reaction rates for several major reactions, like Hydro-demetallization (HDM) and Hydro-desulfurization (HDS), simply because there was limited information on heavy oil components at molecular level. The Petroleomics approach based on molecular level feed and products information and statistical analysis can deal with the heavy oil upgrading process as molecular kinetics models, and then it is anticipated that products could be controlled at the molecular level selectivity as shown in Fig. 1.

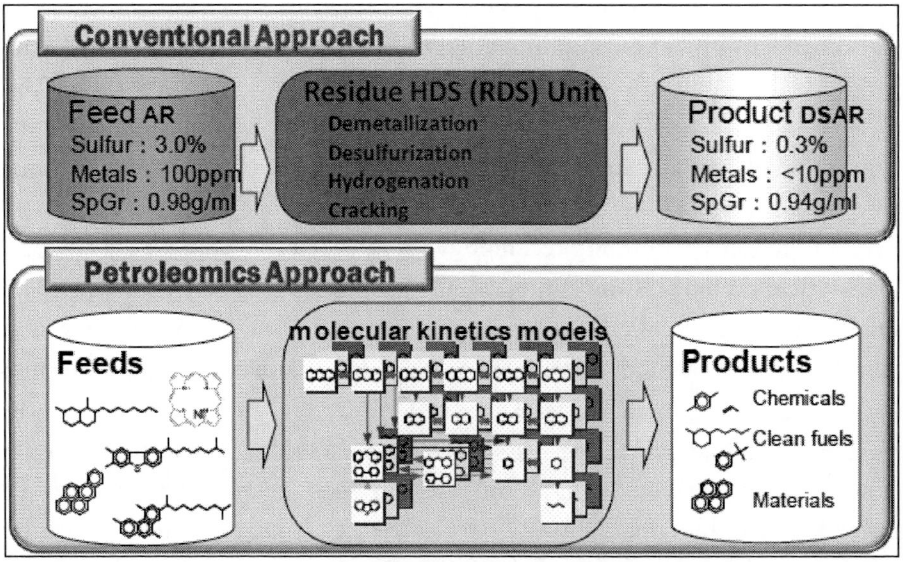

Figure 1. Innovation in heavy oil upgrading process by Petroleomics approach

2. Design and construction of heavy oil component database at molecular level

As a fundamental stage (from FY 2011 to FY 2015), we initiated to build an oil component database at molecular level from structural analyses with FT-ICR MS to the informatics approach on petroleum in combination of each structure and physical properties.

2.1. Analytical procedure

(1) Sample preparation

Heavy oil samples, such as Atmospheric Residue (AR) and Vacuum Residue (VR), were separated into seven fractions by column chromatography: saturates (Sa), one-, two-, three and further multi-rings aromatics (1A, 2A, $3A^{+}$), polar and polyaromatic resins (Po, PA), and asphaltenes (As). The schematic procedures are shown in Fig. 2 [7]. All fractions were dissolved in toluene as 0.5 mg/ml solution before measurements.

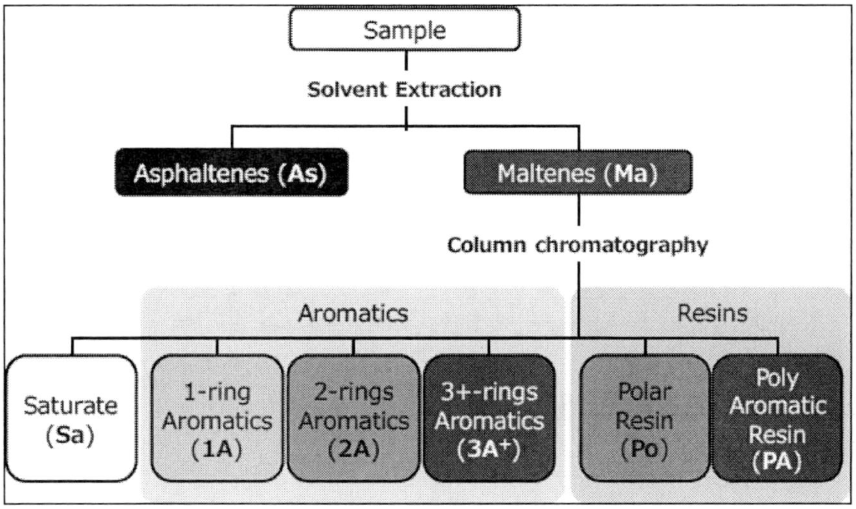

Figure 2. Schematic image of fractionation procedures for heavy oil sample

(2) FT-ICR MS measurements [7, 8]

All mass spectrometer (MS) measurements were performed on FT-ICR MS (solariX 12T, Bruker Japan K.K.) equipped with a 12 T refrigerated actively shielded superconducting magnet. Dissolved fractions of Sa, 1A, 2A, $3A^{+}$, and Po were analyzed using atmospheric pressure photoionization (APPI) in positive ion mode. Nebulizing temperature and flow rate of the sample injection were set at 350°C and 200 μl/h, respectively. Other fractions (PA, As) were ionized by laser desorption ionization1 (LDI) because of poor ionization with APPI. In the measurements with Collision Induced Dissociation (CID), ions passing the quadrupole mass analyzer (Q1) were sent into the collision cell. Q1 is opened and all parent ions were subjected to CID. Collision energy (the voltage difference between Q1 and the collision cell) was set at 30eV to break mainly alkyl side chains and linkages of each petroleum molecules-block (cores). In this condition, parent ions decompose into the constituting core structures, where each core structure is composed of a combination between aromatic, naphthenic and hetero-atom rings, as fragments ions.

(3) Data Analysis

MS peak detection, internal calibration and assignment of elemental formulas were performed with Composer software (Sierra Analytics), and the other procedures were shown in Ref. [7]. A chemical structure of each oil molecule can be defined by a combination between core structures, bridge and side chains, where the order of combination among each component is not considered and these derivatives (isomers) are treated as the same one. The analytical procedure is schematically shown in Fig. 3.

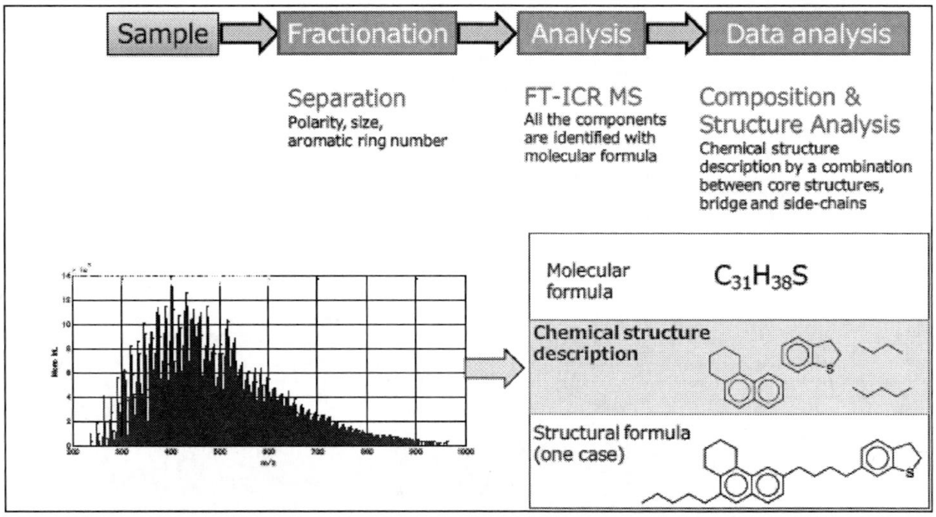

Figure 3. Analytical procedure for chemical structure description

2.2. Petro-Informatics Database

A comprehensive database for petroleum molecules was constructed with information of described chemical structure and physical property. Once the chemical structure is defined, some bulk properties are possible to calculate. Physicochemical properties such as boiling and melting point are obtained by using the group contribution method. Other thermochemical properties such as formation energy and heat capacity are provided from quantum chemistry calculation. Extensive effort to construct the database for all petroleum molecules is ongoing, and the number of registered molecules is up to 25 million.

2.3. Molecular-based reaction modeling for RDS and Asphaltene aggregation

For application of the Petro-Informatics Database as a platform to reaction modeling, residue desulfurization (RDS) was selected at first. Reaction behavior of involved cores was investigated using the chemical structure description data of cores during the RDS reaction. As a result, 1233 types of cores were selected as the dominant covered over 95 mol% in the feed and products, and reaction networks of cores were proposed consisting of 2107 reaction pathways as an example [9].

The other application was Asphaltene aggregation modeling and it is described in the next section.

3. Now we are in the application stage for practical usages in refineries

Now we are in the second stage for the Petroleomics project since FY 2016. Baseed on the developed technologies in the fundamental stage, application and utilization technologies are being developed aiming for practical usages of Petroleomics in refineries, where our developed technologies could shed more light on performance diagnoses, and then to be fulfilled our goals such as operational optimization, and new catalyst & process development, as shown in Fig. 4. Some of main subjects in progress are explained as follows.

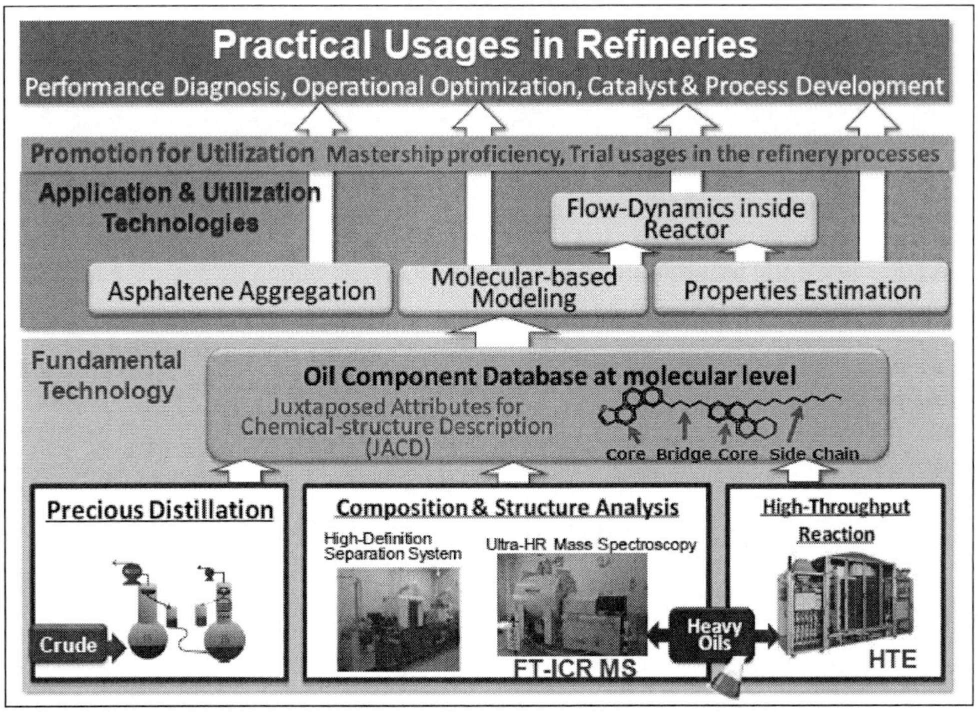

Figure 4. Systematic progress in the Petroleomics technologies

3.1. Crude Oil Database with Petroleomics

Although crude oil sources are preferred to be lighter in Japan at average 36.0° API gravity (S: 1.45 wt%), 2015, a feasibility study for processing heavier crude oils, such as unconventional oils (oil sands) and heavy oils, will be essential, then we initiated to create an original crude oil database with Petroleomics. Selected 12 crude oils have been processed at present and are listed in Fig. 5.

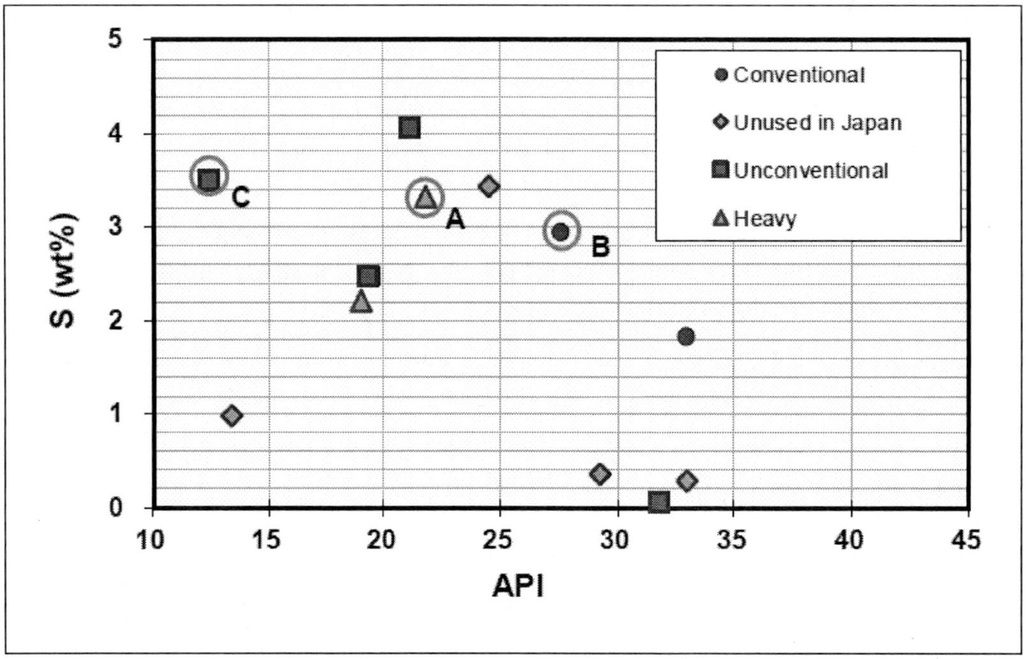

Figure 5. Current lineup of crude oils for database

As an example, number of carbons versus DBE plots as heavy component distribution diagrams, are shown in Fig. 6 for three AR (Atmospheric residues cut at 360°C). DBE (Double Bond Equivalent) values were calculated from chemical formulas using the following equation:

$$\text{DBE} = \text{c} - \frac{\text{h}}{2} + \frac{n}{2} + 1$$

(for the ions $CcHhNnOoSs^{+}$ or $CcHhNnOoSs^{-}$)

For AR of Crude-A, components with higher DBE values at the number of carbons around 35 are higher population than those of Crude-B, indicating that there are more Polycyclic Aromatic Hydrocarbon (PAH) with 6 – 10 aromatic rings and less side chains. While, components are widely distributed from PAH to 1-2 ring aromatic with lots of side chains for AR of Crude-C. Nitrogen containing components in Crude-A and Crude-C are much higher than that in Crude-B (Arabian Heavy as a conventional oil). HDS reactivities of those AR's are now investigated with considering their compositions.

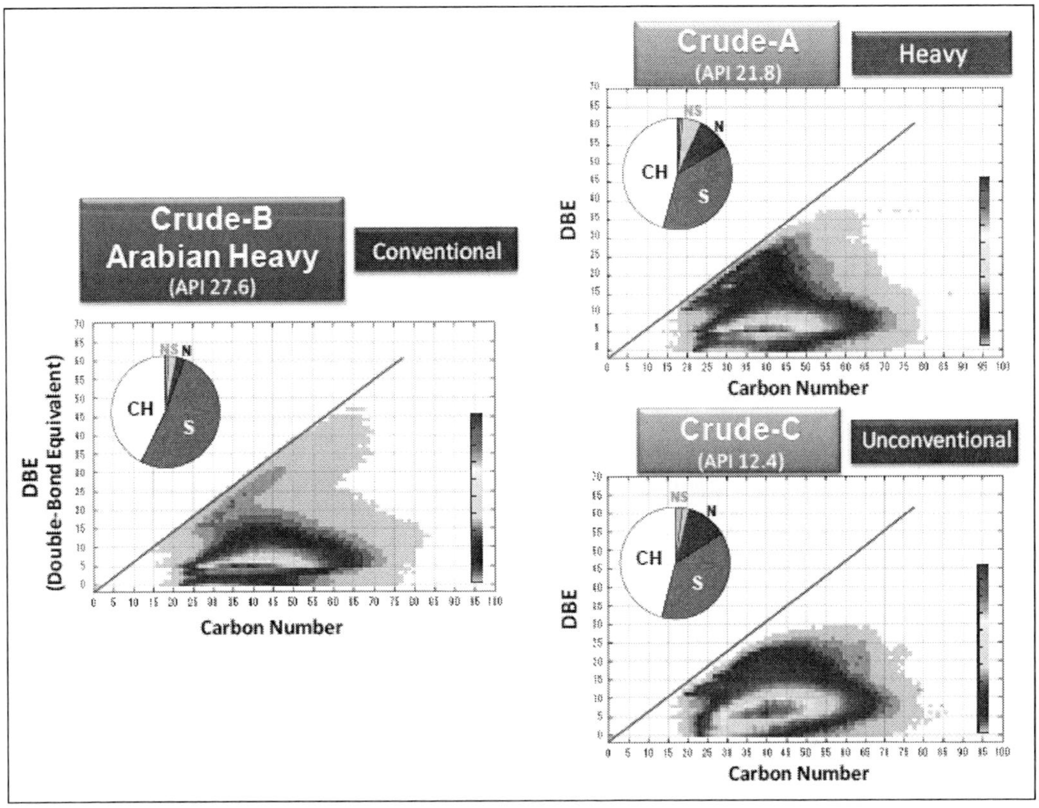

Figure 6. Carbon number – DBE distribution plots for some Atmospheric Residues with composition ratios in pie charts.
(DBE, Double-Bond Equivalent, is explained in text)

3.2. Total optimization of RDS and RFCC operations with molecular-based reaction modeling

As described in the section 2.3, RDS kinetic reaction modeling was developed [9], and subsequently the modeling was applied for simulated evaluation of molar fraction distribution for main core components inside the RDS reactor from the RDS feed to the products (Desulfurized AR: DSAR) . In the next step, the study of molecular-based reaction modeling is being expanded to Residue Fluid Catalytic Cracking (RFCC) process as shown in Fig. 7.

Being applied in the same procedure as for the RDS reactions, most of molecular components in DSAR as a feed of RFCC were chased to their products after the RFCC. Figure 8 shows comparative diagrams for 22 types of core components between products and the RFCC feed, where analyzed cores are covered 98% of whole RFCC products. One and two aromatic-ring cores exist as majority in the products, which are to be made through aromatization and/or de-hydrogenation [10]. On the other hands, PAH cores with more than four aromatic-rings undergo very little change in quantity with the exception of some cores. The investigation is ongoing for total optimization of RDS and RFCC operations.

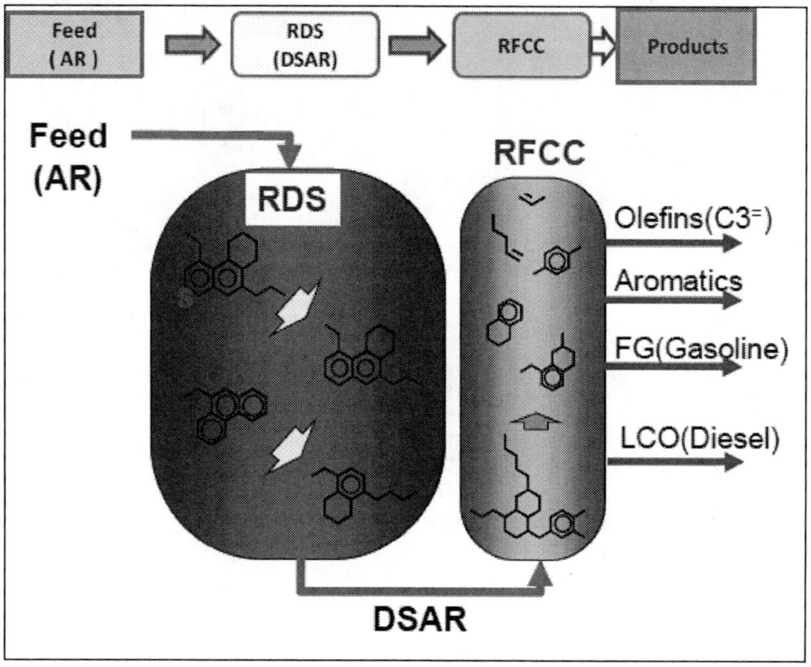

Figure 7. Molecular-based reaction modeling for RDS – RFCC

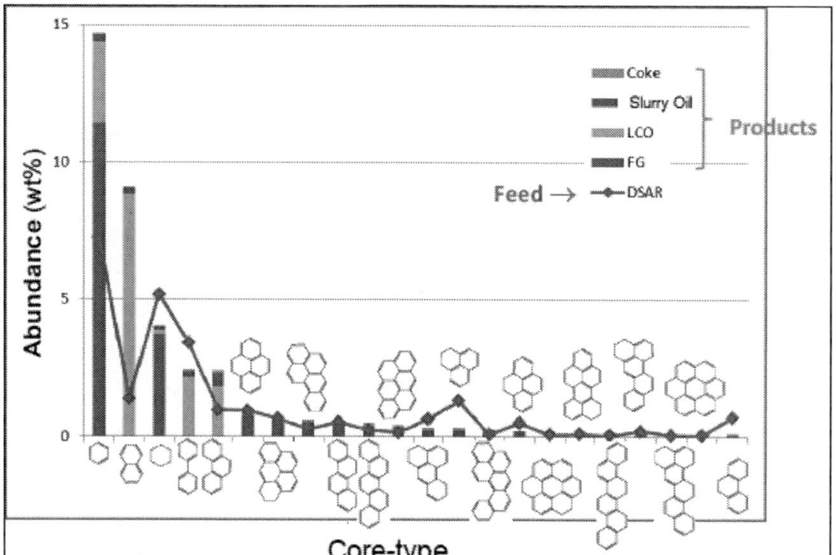

Figure 8. Comparative diagrams for 22 types of core components between products and the RFCC feed

3.3. Mechanism of asphaltene aggregation

The aggregation model for asphaltenes (As) to sediments, which are responsible for the primary causes of fouling and plugging, has been developed as Multi-Component Aggregation Model (MCAM). The model consists of the Hansen solubility parameter (HSP) defined by C.M. Hansen [11] and has been developed by a joint research group between Kansai University (H. Yamamoto), National Institute of Advanced Industrial Science and Technology (AIST) and JPEC [12, 13]. Each of molecular-based heavy oil component in asphaltene can be attributed to a peculiar HSP value from its compositional data. Distribution of each component between soluble, aggregate and precipitate phases is now able to determine by the degree of asphaltene aggregation (D_{agg}) as a function of HSP values (the component and the liquid phase), concentrations and temperature [13].

The developed asphaltene aggregation model is being applied for some refinery tasks, such as the degree of sediment estimation from the feed compositions in some heavy oil upgrading processes, and crude oil stability after mixing (miscibility) in a storage tank. A successful example is a case study of precipitates (sediment) estimation in the heat exchanger after the ebullated-bed Resid hydrocracking unit, as shown in Fig. 9.

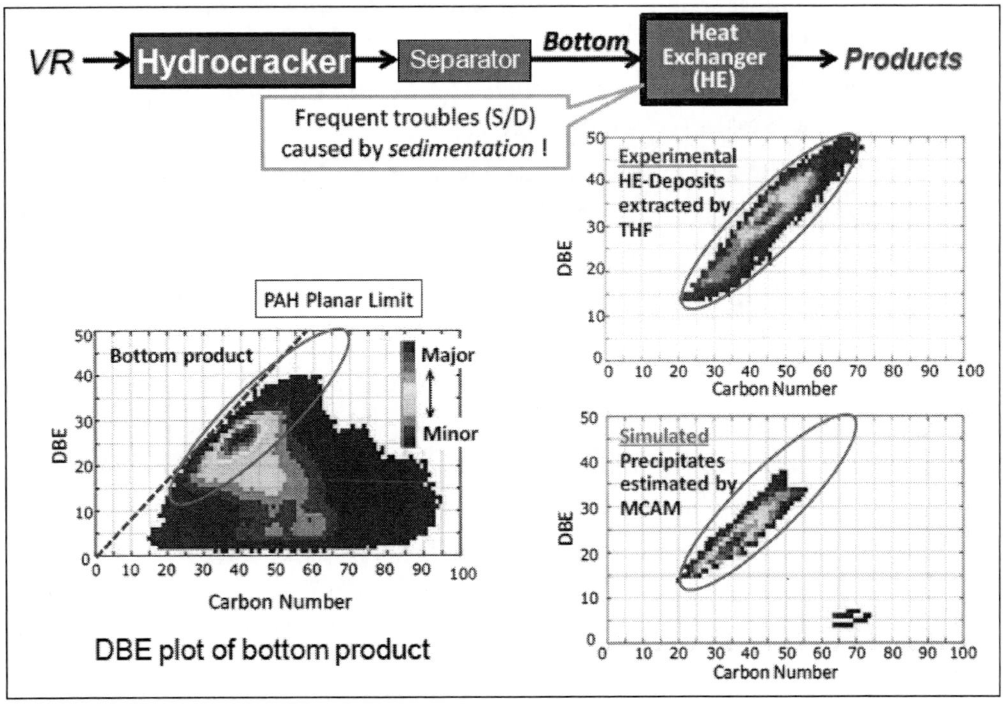

Figure 9. Precipitates estimation for the Resid hydrocracker by MCAM

4. Conclusion

Our Petroleomics project is reviewed from the fundamental stage started in FY 2011 to the ongoing application stage. The crude oil database with Petroleomics is expanding in terms of both quality & quantity, and expected to be an indispensable treasure as a shared infrastructure for dramatic progress in operational reliabilities and productivities in our refineries.

Acknowledgements

This project has been supported Japanese Ministry of Economy, Trade and Industry (METI), and carried out by Japan Petroleum Energy Center (JPEC) and the joint research groups as a technological development project for Petroleomics.

5. References

[1] *GC × GC Handbook, Fundamental Principles of Comprehensive 2D GC*, SHIMADZU, C146-E177 (2012).

[2] Eds. H. Toulhoat, P. Raybaud, *Catalysis by transition metal sulphides*, Editions Technip (2013).

[3] C.S. Hsu, Z. Liang, J.E. Campana, Anal. Chem., **66**, 850 (1994).

[4] R.P. Rodgers, A.G. Marshall, *Asphaltenes, Heavy Oils and Petroleomics*, Springer, New York, pp. 63-89 (2007).

[5] A.G. Marshall, R.P. Rodgers, *Petroleomics: Chemistry of the underworld*, Proc. Natl. Acad. Sci. USA, **105** (47), 18090 (2008).

[6] C.S. Hsu, C.L. Hendrickson, R.P. Rodgers, A.M. McKenna, A.G. Marshall, *Petroleomics: Advanced molecular probe for petroleum heavy ends*, J. Mass Spectrom. **46**, 337 (2011).

[7] K. Katano, T. Suzuki, R. Tanaka, Prep. Pap.-Am. Chem. Soc., Div. Energy and Fuels, **60** (2), 532 (2015).

[8] K. Katano, T. Suzuki, R. Tanaka, Prep. Pap.-Am. Chem. Soc., Div. Energy and Fuels, **61** (2), 694 (2016).

[9] K. Hagiwara, T. Gondai, H. Fujinaga, R. Tanaka, J. Japan Petro. Inst., **59** (5), 219 (2016).

[10] I. Shimada, K. Takizawa, H. Fukunaga, N. Takahashi, T. Takatsuka, Fuel, **161**, 207 (2015).

[11] C.M. Hansen, *Hansen Solubility Parameters: A User's Handbook, 2nd ed.*, CRC Press (Boca Raton, FL) 2007.

[12] T. Sato, S. Araki, M. Morimoto, R. Tanaka, H. Yamamoto, Energy & Fuels, **28**, 891 (2014).

[13] M. Morimoto, T. Sato, S. Araki, R. Tanaka, H. Yamamoto, S. Sato, T. Takanohashi, Energy & Fuels, **29**, 2808 (2015).

PAPER 12

Effect of alumina binder on the deactivation behavior of Pt-Beta catalyst for transalkylation of toluene with 1,2,4-trimethylbenzene

Faisal Almulla
Saudi Aramco R&DC

Benzene, toluene, and xylene are three basic raw materials for the production of most aromatic derivatives such as polyesters, plastics and detergents. Xylenes (*p*-, *m*- and *o*-) have the greatest market demand with an increasing annual rate of 6%. Transalkylation process is of increasing importance to convert the surplus toluene and low value of C_9 aromatics into more valued products, such as xylenes. Catalyst deactivation remains a major challenge in transalkylation process.

Using industrially relevant operating parameters, the transalkylation of 1,2,4-trimethylbenzene (TMB) with toluene was studied over zeolite Beta. A fixed bed reactor was operated at WHSV of 5 h^{-1}, 400 °C, 10 bar and a 50:50 wt. % toluene:TMB ratio where the xylenes yield found to be 25 wt. % after 50 h time-on-stream (TOS).

Incorporation of Pt (0.08 wt. %) further improved the activity of the catalyst and increased xylenes yield to 40 wt. %. Embedding the catalyst with certain levels of gamma alumina as a binder was carried out to reduce the catalytic cost and enhance the mechanical strength of the catalyst. The Pt-Beta/alumina mixture had minor impact in reducing the activity of the catalyst with increasing amount of binder. The optimum binder content was shown to be 40 wt. %. This catalytic composition (60;40%, PtBeta;Al) was chosen because it yielded the same amount of xylenes as when 20 wt. % binder was used.

PAPER 13

Driving catalysis innovation in petrochemicals through designing unneglectable support materials

Dr. Muhammad H. Haider, Yasser Turki Al-Harbi, Chandrasekar Subramani.*
SABIC Technology Center, Riyadh KSA 11551. HaiderMh@sabic.com

Competitive petrochemical technologies are now prime focus in the current era of uncertain global trends in petrochemicals economies. The main area being explored to next level is the development of certain catalytic materials which are capable of outshining the traditional less robust catalysts. The modern catalysts have better conversion selectivity and TOS features which can lead to massive downstream energy saving and in turn less cost per ton production. Active phases on the catalysts have been thoroughly looked at by the scientific community around the world. However, the support material which plays crucial role still requires attention. Here in this talk, the new features of thoroughly designed support material applicable to diverse PetChem technologies, for example oxygenates, olefins, di-olefins production etc., are presented. It is strongly believed that support material, if properly designed and produced, can give the desired

PAPER 14

Study the nano-structured zeolites for hydrocarbon reforming: kinetic and reaction pathways

Ali N. AlJishi [a], Emad N. Al-Shafei [a], Oki Muraza [b], Ki-Hyouk Choi [a] and Ali S. Alnaser [a]

a) Research and Development Center, Saudi Aramco, Dhahran 31311, Saudi Arabia
b) Center of Excellence in Nanotechnology and Chemical Engineering, King Fahd University of Petroleum & Minerals (KFUPM), Dhahran 31261, Saudi Arabia

Zeolites are the potential catalysts for refining and petrochemical industry applications. This study is conducted on naphtha reforming process to produce higher octane and aromatic rich reformate to extract BTX for Petrochemicals. The synthesized nanocrystal zeolite catalysts were characterized by ammonia temperature programmed desorption (NH_3-TPD), XRD, NMR, and N_2 physisorption. The catalytic testing of hydrocarbon cracking was carried out in a fixed bed reactor at a space velocity of 10 h^{-1} to investigate the reaction pathway. In zeolitic reactions, there are two main pathways, which might occur: primary cracking and secondary cracking. Those mechanisms could be resulted from the micropores of the zeolites where the reaction takes place. The developed lumping kinetics of dodecane cracking involved n-paraffins, iso-paraffins, olefins, napthenes, and aromatics. They were described via fourteen rate constants in order to recognize the key reaction pathways and draw mechanism of short-path length pores of nano zeolite. The presentation will discuss the kinetic modeling of dodecane cracking over nano zeolite catalysts to predict products distribution as well as estimated rate constants parameters of hydrocarbon reforming.

PAPER 15

Advance characterization of heterogeneous catalyst of core shell Pt@SiO2 nanoparticles

Dr. Noor Al-Mana
Research & Development Center, Saudi Aramco, Dhahran, Saudi Arabia

A heterogeneous catalyst is one of the mostly deployed/used materials in industrial processes. A good catalyst's key features include: long-lasting, highly selective and active, low-cost, robust, and environmentally benign. In this context, a deep understanding of the correlations between catalyst structures and their properties is essential for the heterogeneous catalysts research.

The core-shell design has attracted a lot of attention due to their distinct structure where the oxide layer encapsulates the metal active site. The protective oxide layer suppresses catalyst deactivation, which is attributable to metal growth (sintering) at elevated temperatures, as it acts as a protective shield to metal migration. Also the unique design might enhance the activity of the catalysts, as new active sites are created at the interfaces between the metal and oxide, especially oxide with redox properties as TiO_2 and CeO_2. Controlling the thickness of this oxide shell is crucial for gases and reactant diffusion through it to reach the active metal site.

To achieve the structure with protected metal nanoparticles, reverse micro-emulsion system was chosen to give metal@oxide core-shell materials. This method allows synthesizing fine and monodispersed metal nanoparticles, followed by control coating of metal nanoparticles by oxides matrix. The reverse micro-emulsion method consists of forming nano-sized droplets of water as a nano-reactor in a mixture of organic solvent (cyclohexane) and surfactant (IGEPAL CO-520). To precisely control the size and thickness of the metal@oxide nanoparticles, different parameters play a role in this synthesis, such as the surfactant and organic solvent nature, pH, water/surfactant ratio and reducing agent. The synthesis of Pt@SiO_2 core-shell particles has been extensively studied.

In this study, advanced analytical methods, such as high resolution transmission electron microscopy (HRTEM), CO-Diffuse Reflectance for Infrared Fourier Transform Spectroscopy (CO-DRIFT), and N_2 physical sorption, are applied for catalyst characterizations. Analytical results provide critical information in understanding the CO diffusion through the SiO_2 layer in the Pt-SiO_2 core-shell structure, which helps the establishment of the relationship between the catalytic structure and superior catalytic activity. Additionally, the stabilization of metal growth by the core-shell system is illustrated by HRTEM images. The mechanism and key parameters to obtain core-shell nanoparticles of Pt and SiO_2 using the reverse micro-emulsion method is also discussed.

PAPER 16

Modelling of the Dehydrogenation Reaction of Alkanes to Alkenes in a Fixed Bed Reactor System

TAREK J. JAMALEDDINE* AND JORGE JIMENEZ†

* Saudi Basic Industries Corporation (SABIC)
SPADC - King Saud University Campus
Riyadh Techno Valley, Riyadh, KSA 11551
E-mail: jamaleddine@sabic.com - web page: http://www.sabic.com

† Saudi Basic Industries Corporation (SABIC)
SABIC - AMERICAS
1600 Industrial Blvd, Sugar Land, TX, US 77478
Email: jjimenez@sabic.com - web page: http://www.sabic.com

ABSTRACT

The catalytic dehydrogenation reaction of alkanes to alkenes is simulated in a heterogenous fixed bed reactor (FBR). The porous medium model in ANSYS Fluent combined with the reaction model capability was utilized to predict the flow behavior and species transport in a bed of spherical particles. Physical and material properties of a dehydrogenating catalyst of Chromium Oxide (Cr_2O_3) on Aluminum Oxide Support (Al_2O_3) were employed in the model. Several reaction models were implemented using a customized User-defined Function (UDF) subroutine. Simulation results were validated against plant data in the literature for a similar process. Good agreement was observed for the conversion of alkanes to alkenes within acceptable accuracy. It is concluded that the Power Law reaction model showed the least fit for the feed conversion and product selectivity compared to the other studied reaction models.

1. INTRODUCTION

The catalytic dehydrogenation process is greatly used in the petrochemical industry to convert Alkanes to Alkenes. The process is performed in a FBR at sub-atmospheric pressure conditions and temperature ranging from 540 – 670 °C. In these reactors, reaction takes place on the surface of catalyst pellets placed randomly inside the reactor volume. Chromium Oxide (Cr_2O_3) based catalyst is generally used and is supported on Aluminum Oxide (Al_2O_3) metal. Side reactions and by-product formation are some of the many challenges characterizing this process. It is thus beneficial to have a profound understanding of the flow characteristics and reaction mechanisms in order to gain more insight into improving the system profitability.

Many researchers in the field have studied the dehydrogenation process. The most relevant to this study is the pioneering work of Airaksinen et al. [1] who presented several reaction mechanisms and kinetic models specifically used for describing the rates of dehydrogenation reaction. Al-Ghamdi et al. [2] presented a paper showing actual plant data for one of the largest dehydrogenation unit for MTBE production. Sanfilippo [3] discussed the current technology for the production of MTBE from isobutylene and methanol, and shed light into the type of catalyst widely used in this process. Gao et al. [4] and Chen et al. [5] developed CFD models that describe the flow behavior and dehydrogenation reaction in a FBR. In their models, they followed a one-way forward reaction and showed that the space velocity had less effect on the reaction outcome and a greater impact on the generation of hot spots, whereas inlet gas temperature had a significant impact on both, the reaction outcome and the generation of hot spots in the bed.

This paper presents a three-dimensional CFD model for the dehydrogenation reaction of alkane to alkene, specifically, isobutane to isobutylene in a FBR. A pilot-scale model was initially studied followed by a scaled-up commercial model. Three reaction models were adopted in this study. These models represent the

*To whom correspondence should be addressed. E-mail address: jamaleddine@sabic.com.

least suitable, the more suitable, and the best suitable with respect to capturing the dehydrogenation reaction of isobutane to isobutylene. Model predictions for conversion and selectivity were validated against tabulated values in Al-Ghamdi et al. [2] of 53 wt% and 93.6 wt%, respectively. Forward and reverse reactions were both considered in this study to accurately capture the reaction mechanism. User-defined function (UDF) subroutines were introduced to enhance Fluents solver capability in simulating reaction rates accurately.

2. CFD MODEL DEVELOPMENT

Figure 1 depicts a schematic of the pilot-scale model for CFD analysis. Due to symmetry, only one-fourth of the actual cylindrical gemoetry was used in the simulation. In the case of a commercial-scale, a slightly different geometry was developed with inlet and outlet zones stretched to actual height of the commercial-scale reactor. The catalyst was assumed to be a continuous porous medium with additional source terms in the governing equations to reflect flow resistances and reaction mechanisms.

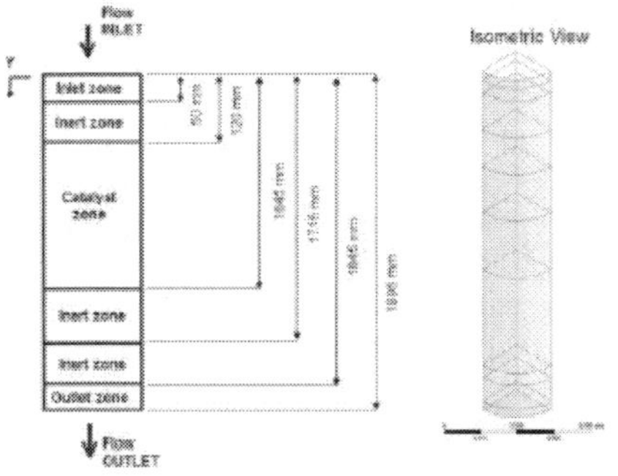

Figure 1. Schematic of the CFD model geometry

The model was primarily discretized with hexahedral elements in regions aligned with the direction of the flow. Tetrahedral and wedge elements were implemented elsewhere. An optimized element count of 1 515 618 elements was used for the pilot-scale, whereas 2 805 518 elements were adopted for the commercial-scale. The *global element* quality was reported to be greater than 0.12 and the *global cell* skewness less than 0.96. Finer mesh was adopted close to the reactor wall (Figure 2) to adequately resolve the boundary layer profile in that region. A grid independency study was also performed (not shown in this paper). Consequently, the grid size in this paper represents the optimum mesh-size needed for an accurate and converged solution based on predicted pressure drop values in the catalyst bed.

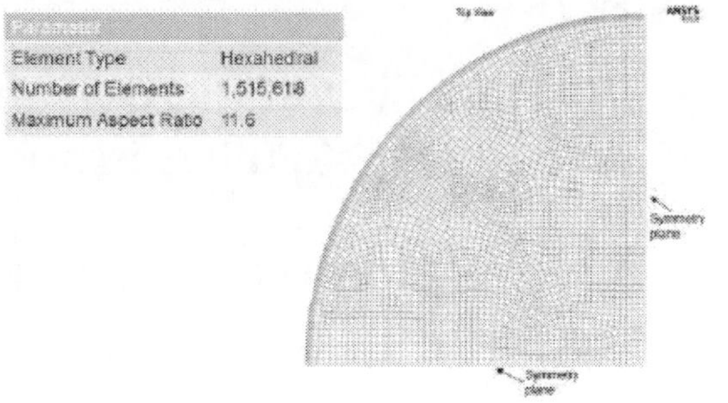

Figure 2. Schematic of the adopted mesh, top view

2.1 Governing Equations

The governing equations were solved using the pressure-based solver in ANSYS Fluent v18. In addition to the continuity, momentum, and energy equations, the species transport equation was also adopted. These equations are given as follows:

Continuity equation

$$\frac{\partial \rho}{\partial t} + \nabla \cdot (\rho \vec{v}) = S_m \quad (1)$$

Momentum balance equation

$$\frac{\partial}{\partial t}(\rho \vec{v}) + \nabla \cdot (\rho \vec{v} \vec{v}) = -\nabla p + \nabla \cdot (\bar{\bar{\tau}}) + \rho \vec{g} + \vec{F} \quad (2)$$

Energy balance equation

$$\frac{\partial}{\partial t}(\rho E) + \nabla \cdot \left(\vec{v}(\rho E + p)\right) =$$

$$\nabla \cdot \left[k_{eff} \nabla T - \textstyle\sum_j h_j \vec{J}_j + \left(\bar{\bar{\tau}}_{eff} \vec{v}\right)\right] + S_h \quad (3)$$

Species transport equation

$$\frac{\partial}{\partial t}(\rho Y_i) + \nabla \cdot (\rho \vec{v} Y_i) = -\nabla \cdot \vec{J}_i + R_i + S_i \quad (4)$$

At the inlet of the reactor, mixture properties were derived based on the ideal gas mixing law and an inlet temperature of 580 °C (Table 1). The feed was asummed to be primarily 99.4 wt% gaseous isobutane to satisfy the actual inlet conditions of the reactor. At the reactor outlet, a pressure outlet condition was specified. A no-slip boundary condition with an adiabatic wall temperature was specified at the reactor walls, and a continuous boundary condition (zero normal gradients of all variables) was specified at the symmetry planes. The porous regions of the reactor were treated as laminar, whereas k-ε turbulent model was assumed elsewhere. A UDF describing the reaction rates was interpreted into the solver and hooked to its respective cell zone dialog box. It took nearly 5 days of wall-clock time for an 8-minutes transient solution to converge on an 8-Core 15-node Intel processor with residual RMS values below 10^{-4}

for all equations except for the energy and turbulent equations with values below 10^{-6}.

Description	Value
Gas mixture	
$C_{p,g}$ (kJ/kmol.K)	Ideal gas mixing law
ρ_g (kg/m^3)	Ideal gas mixing law
μ_g (Pa.s)	Ideal gas mixing law
k_g (W/m.K)	Ideal gas mixing law
Solid phase	
ρ_g (kg/m^3)	1520
d_p (m)	1.98 x 10^{-2}
Porosity, ϕ	0.37
Other model parameters	
Space velocity (h^{-1})	495.068
Feed gas temperature (K)	853.15

Table 1. Model parameters

2.2 Reaction Kinetics Model

The dehydrogenation reaction of isobutane is completed in the presence of solid catalysts and includes both forward and reverse reaction rates as shown in Equation (1).

$$i - C_4H_{10} \leftrightarrow i - C_4H_8 + H_2 \tag{1}$$

Various types of kinetic models are available in the literature [6-11]. These models are derived based on possible dehydrogenation mechanisms and reaction rate-determining steps. It was concluded in Airaksinen et al. [1] that desorption models described the activity results poorly, whereas models describing adsorption and/or surface reaction as the rate-determining step gave the best fit to experimental data. Consequently, desorption models were not considered in this study.

Three kinetic models were assessed and validated to achieve the best model that closely fit the reported conversion and selectivity rates. These models are discussed below including parameters and reaction constants derived from experimental observations.

Power Law Model

This model is presented in Equation (2) and is based on Arrhenius expression which assumes that the rate constant of a chemical reaction is dependent on the absolute temperature as shown in Equation (3). It neglects turbulent fluctuations and side reactions such as cracking and coke formation. In addition, the model is widely used for homogenous reactions.

$$-r_A = k\left(p_A - \frac{p_E p_{H_2}}{K}\right) \tag{2}$$

$$k = A_{mean} e^{(-E/R)(1/T - 1/T_{mean})} \tag{3}$$

Such that parameters for this model are:

Pre-exponential factor, $A_{mean} = 0.0313 \frac{mol}{(kgcat.s.bar)}$

Activation energy, $E = 99\ kJ/mol$

Eq. constant, $K = 2.1 \times 10^7 \exp\left(-\frac{122\ kJ/mol}{RT}\right)\ bar$

$T_{mean} = 550$ °C

Langmuir-Hinshelwood Model

This model describes the dehydrogenation reaction based on the principles of the Langmuir-Hinshelwood approach [12]. It is widely used for heterogeneous reactions with the reaction rate given as:

$$-r_A = \frac{k'\left(p_A - \frac{p_E p_{H_2}}{K}\right)}{1 + K_{EH2} p_E p_{H2} + K' p_E p_{H2}^{1/2} + (K_{H2} p_{H2})^{1/2}} \tag{4}$$

Such that the parameters for this model are:

Pre-exponential factor, $A_{mean} = 0.0596 \frac{mol}{(kgcat.s.bar)}$

Activation energy, $E = 141\ kJ/mol$

Eq. constant, $K = 2.1 \times 10^7 \exp\left(-\frac{122\ kJ/mol}{RT}\right)\ bar$

$K_{EH2} = 8 \times 10^{-6}\ bar^{-2}$

$K' = 22.9\ bar^{-3/2}$

$K_{H2} = 7.56\ bar^{-1}$

Happel et al. Model

This model [13] is a simpler form of the Langmuir-Hinshelwood's. It assumes adsorption as the rate-determining step and the reaction rate takes the following form:

$$-r_A = \frac{k'\left(p_A - \frac{p_E p_{H_2}}{K}\right)}{1 + K_E p_E + K_{H2} p_{H2}} \tag{5}$$

Such that the parameter estimation for this model are:

Pre-exponential factor, $A_{mean} = 0.0557 \frac{mol}{(kgcat.s.bar)}$

Activation energy, $E = 142\ kJ/mol$

Eq. constant, $K = 2.1 \times 10^7 \exp\left(-\frac{122\ kJ/mol}{RT}\right)\ bar$

$K_E = 8.76\ bar^{-1}$

$K_{H2} = 4.28\ bar^{-1}$

The temperature dependence of the reaction rate constant was described with an Arrhenius equation rewritten as:

$$k = A_{mean} e^{-zE} \tag{6}$$

where $A_{mean} = A_0 e^{-E/_{RTmean}}$

$z = 1/R \ (1/T - 1/Tmean)$

and a temperature of 550 °C was assumed for T_{mean}.

The equilibrium constant of the dehydrogenation reaction also taken from Airaksinen et al. [1]:

$$K = 2.1 \, x \, 10^7 exp\left(\frac{-122 \, kJ/mol}{RT}\right) \text{bar} \tag{7}$$

Since the default reaction formulation in Fluent solver has the ability to solve the Power Law model only, UDF was implemented to test the heterogeneous reaction expressions given by the Langmuir-Hinshelwood [12] and Happel et al. [13] models. For additional details about the reaction mechanisms, the reader is referred to Airaksinen et al. [1].

The customized macro functions were introduced into the solver as source terms in the energy and species transport equations. In order to validate the feasibility of the UDF, a preliminary simulation was conducted on the pilot-scale model with the default volumetric reaction (Power Law formulation) of the Fluent solver. Results from this simulation were then compared with those obtained using an in-house UDF in the absence of volumetric reaction.

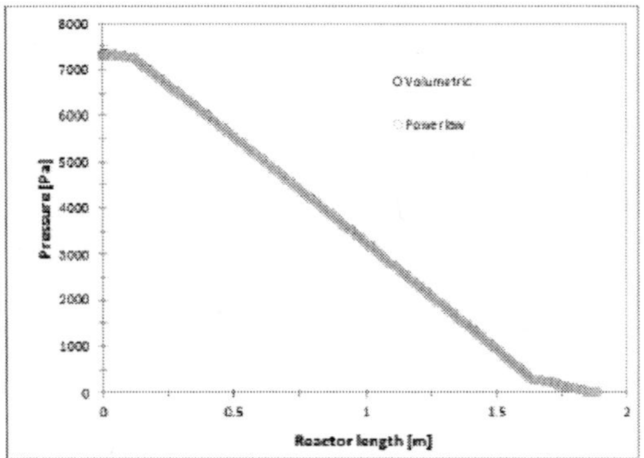

Figure 3: Pressure drop along the reactor length

As can be seen in Figures 3 and 4, the predicted pressure drop and product selectivity matched perfectly well for both simulations across the reactor height. This implies the successful coding and implementation of the customized UDF. In order to implement the aforementioned heterogenous reactions, the successful UDF was subsequently altered to reflect Equations (4) and (5). Theoretical conversions and selectivities were calculated on w.t.% basis as shown in Equations (8-9).

$$Conversion = 100\% \times \frac{Y_{in} - Y_{out}}{Y_{in}} \tag{8}$$

$$Selectivity = 100\% \times \frac{Y_i}{Y_{in} - Y_{out}} \tag{9}$$

where Y_{in} and Y_{out} denote the mass fraction of isobutane at inlet and outlet, respectively and Y_i denotes the mass fraction of isobutylene at any reactor location. The Yield of isobutylene was calculated as the product of conversion and selectivity.

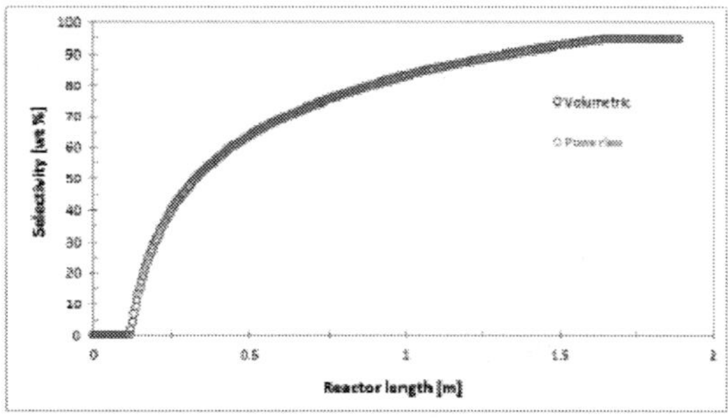

Figure 4: Product selectivity along the reactor length

3. RESULTS

3.1 Pilot-Scale Model

Figure 5 shows a contour plot of the mass fractions of isobutane and isobutylene along the length of the reactor. As can be seen in the figure, the mass fraction of isobutane fell sharply in the early stages of reaction coupled with a corresponding increase in isobutylene formation. This indicates the consumption of isobutane to generate isobutylene. It can also be seen that slower reaction rates occurred near the end of the catalyst region; an indication of the presence of redundant catalyst in that region. These results also confirm the success of the CFD model in capturing the reaction mechanism reasonably well.

Figure 6 shows the theoretical predictions of conversion and selectivity for the considered kinetic models. It can be drawn from this figure that Happel et al. model [13] showed the closest fit for the isobutane conversion to isobutylene within an error of 5% in comparison to the tabulated conversion data shown in Al-Ghamdi et al. [2]. The Power Law model underestimated the conversion rate by nearly 54%. The discrepancy between theoretical and tabulated conversion data can be attributed to the values of the kinetic parameters used in the theoretical model. In addition, Al-Ghamdi et al. [2] reported higher operating temperatures in the range of 580 to 620 °C thus triggering higher conversion rates.

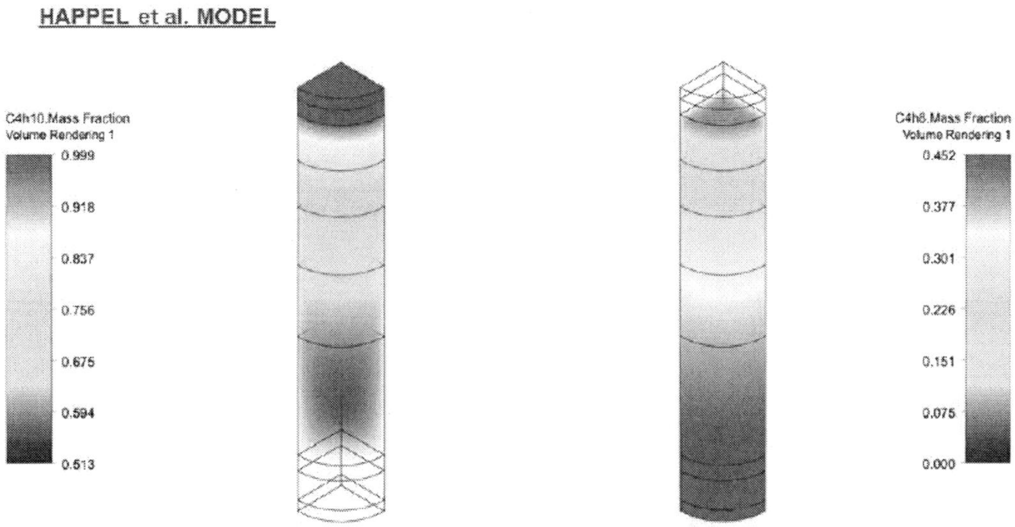

Figure 5. Mass fraction of reaction components

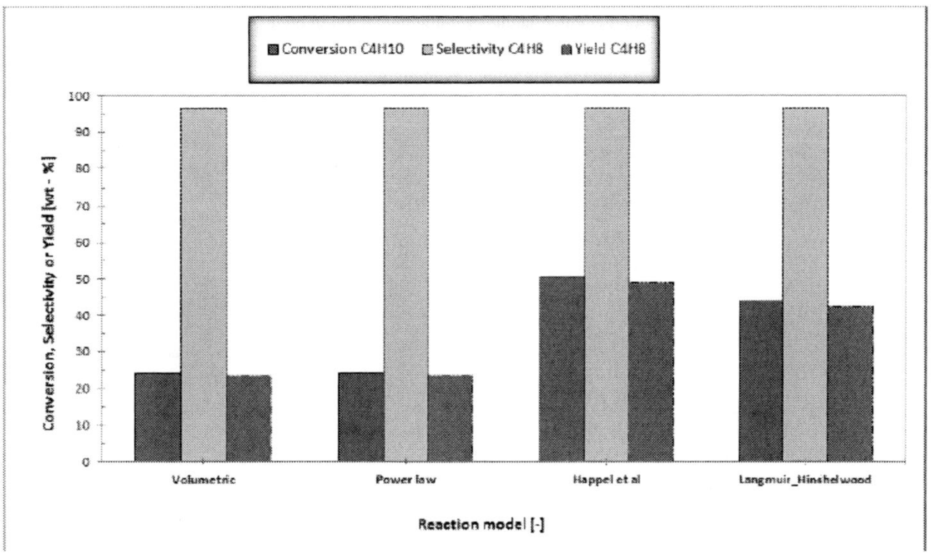

Figure 6. Theoretical prediction

3.2 Commercial-Scale Model

The success of the pilot-scale model triggered an interest to test the code on a commercial-scale similar to the one shown in Al-Ghamdi et al. [2]. CFD predictions for the isobutane conversion and isobutylene selectivity are presented in Figure 7. It can be clearly seen in the figure that the Power Law model underestimated conversion rates as shown in the previous section with respect to Al-Ghamdi et al. [2] value by 54%. The other reaction models however showed a better agreement. This indicates that the nature of the reaction is not homogenous but rather heterogeneous as expected. The figure also showed a better agreement of Happel et al. [13]

*To whom correspondence should be addressed. E-mail address: jamaleddine@sabic.com.

model with the published data then Langmuir-Hinshelwood's [12]. The later approach predicted a slightly lower consumption of isobutane to form isobutylene; however, both models predicted the conversion and selectivity rates reasonably well. The overestimation of selectivity rates by 4% can be attributed to the absence of coking and side reactions in the given models that tend to rather decrease the product selectivity.

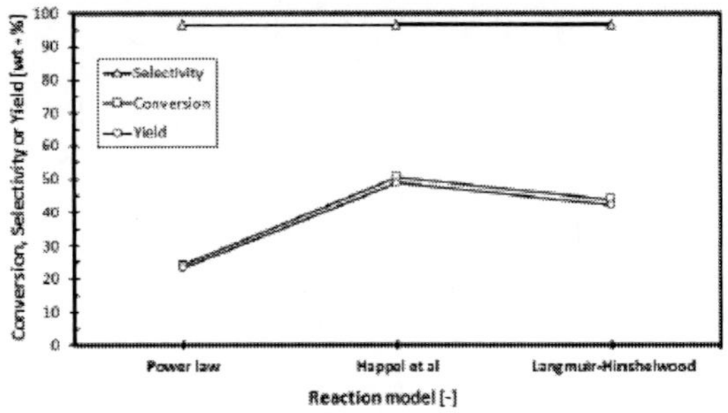

Figure 7. CFD prediction for conversion and selectivity

Figure 8 shows the predicted reaction rates in the catalyst bed. A steep drop followed by slow reaction rates can be clearly seen in the figure for the successful model given by Happel et al. [13]. This indicates the fast and spontaneous nature of the dehydrogenation reaction of isobutane to isobutylene. The impact of the reverse reaction on the overall reaction rate was fairly negligible; consequently, it was not included in the plot. Furthermore, the presence of a nearly flat profile close to the mid-height of the bed and beyond indicates redundancy in the amount of catalyst in that region and a potential cost saving on that part.

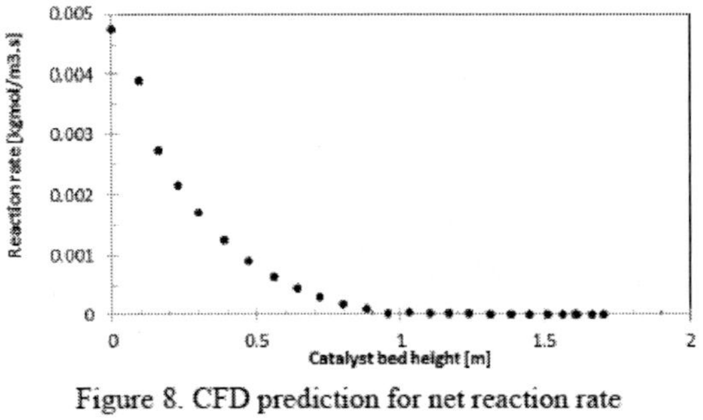

Figure 8. CFD prediction for net reaction rate

Figure 9 depicts the bed temperature along the height of the bed. It is anticipated to see a drop in temperature due to the nature of the dehydrogenation reaction. The model predicted a 35 °C drop thus confirming the endothermic nature of the reaction and the success of the customized reaction model. It is worth to mention that the temperature peak at the bed inlet is due to an imposed temperature profile on the model as an intial bed condition due to regeneration cycle ahead of dehydrogenation process.

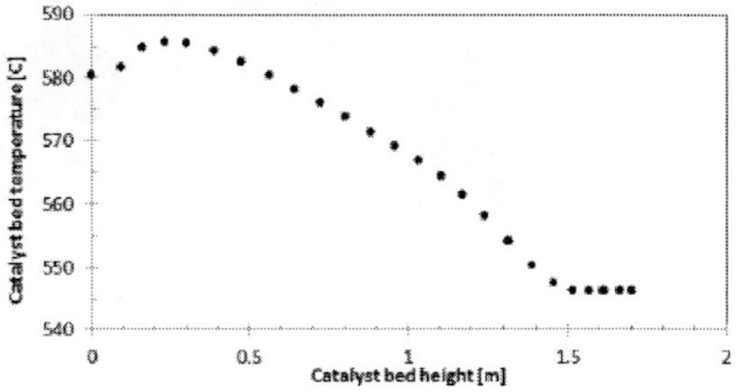

Figure 9. CFD prediction for bed temperature distribution

4. CONCLUSION

CFD analyses were performed on a pilot-scale and a commercial-scale reaction model of a fixed bed CATOFIN® dehydrogenation reactor to predict the reaction kinetics and product yield. Several kinetic models were examined based on proposed reaction mechanisms in the open literature for a similar process. CFD predictions were validated against published plant data for isobutane conversion rate and isobutylene selectivity. It was concluded that:

1. The customized CFD model was successful in predicting isobutane conversion rate. The kinetic model proposed by Happel et al. [13] showed the closest fit within an error of 5%.
2. The customized CFD model was successful in predicting isobutylene selectivity within an error of 4% when the kinetic model proposed by Happel et al. [13] was adopted.
3. The Power Law model underestimated the conversion rate of isobutane by 54% and overestimated the selectivity by 6%.
4. All models over predicted the selectivity to isobutylene within an error of 10%.
5. Slower reaction rates were observed at the end of the catalyst zone. This denotes the presence of a redundant catalyst region in the bed and a potential cost saving on that part.

Nomenclature

A_{mean}, A_o = Pre-exponential factor, mol/(kgcat.s.bar)

d_p = Particle diameter, m

E = activation energy, KJ/mol

k_g= Thermal conductivity, W/m.k

k'= apparent rate constant, mol/(kgcat.s.bar)

K = reaction equilibrium constant, bar

K_i = adsorption equilibrium constant of species i, bar^{-1}

K_{EH2}, K' = combined equilibrium constants

pi = partial pressure of species i, bar

ri = reaction rate of component i, mol/(kgcat s)

R = universal Gas constant, 8.314 J/(mol K)

T = temperature, K or °C

ρ = density, kg/m^3

T = time, s

v = velocity, m/s

S_m = source term

S_h = heat of chemical reaction

Y_i = mass fraction of species i

R_i = net rate of production of species

S_i = rate of creation

μ_g = dynamic viscosity, (Pa.s)

Subscripts

A = isobutane

E = isobutene or isobutylene

H2 = hydrogen

EH2 = isobutylene and hydrogen

g = gas phase

s = solid phase

5. References

[1] Airaksinen, S.M.K., Harlin, M.E., and Krause, A., 2002, "Kinetic modelling of dehydrogenation of isobutane on chromia/alumina catalyst," Ind. Eng. Chem. Res., 41, pp. 5619-5626.

[2] Al-Ghamdi, A.A., Al-Bassam, A.M., and Draus, J.L., 1994, "Start-up of World's largest Isobutane dehydrogenation plant for MTBE production," Annual Meeting of the National Petroleum Refiners Association, San Antonio, TX, paper AM-94-28, pp. 1-12.

[3] Sanfilippo, D., 2011, "Dehydrogenations in fluidized bed: Catalysis and reactor engineering, Catalysis Today, 178, pp. 142-150.

[4] Gao, Xi., Zhu, Y., and Luo, Z., 2011, "CFD modeling of gas flow in porous medium and catalytic coupling reaction from carbon monoxide to diethyl oxalate in fixed-bed reactors," Chemical Engineering Science, 66, pp. 6028-6038.

[5] Chen, X., Dai, J., and Luo, Z., 2013, "CFD modelling using heterogeneous reaction kinetics for catalytic dehydrogenation syngas reactions in a fixed-bed reactor," Particuology, 11, pp. 703-714.

[6] Dodd, R. H., and Watson, K. M., 1946, "Process design of catalytic reactors – Dehydrogenation of Butane," Trans. Inst. Chem. Eng., 42, pp. 263-291.

[7] Carra, S., Forni, L., and Vintani, C., 1967, "Kinetics and mechanism in catalytic dehydrogenation of n-Butane over Chromium-Alumina", J. Catalysis, 9(2), pp. 154-165.

[8] Happel, J., and Atkins, R. S., 1970, "Rate-controlling path for catalytic butane dehydrogenation," Industrial and Engineering Chemistry Fundamentals, 9(1), pp. 11-14.

[9] Zwahlen, A. G., and Agnew, J. B., 1992, "Isobutane dehydrogenation kinetics determination in a modified Berty gradientless reactor," Industrial and Engineering Chemistry Research, 31(9), pp. 2088-2093.

[10] Happel, J., Kamholz, K., Walsh, D., and Strangio, V., 1973, "Kinetics of the Isobutane-Isobutene-Hydrogen system using tracers," Industrial and Engineering Chemistry Fundamentals, 12(3), pp. 263-267.

[11] Jackson, S., Grenfell, J., Matheson, I. M., and Webb, G., 1999, "Modelling of alkane dehydrogenation under transient and steady state conditions over a chromia catalyst using isotopic labelling," Studies in Surface Science and Catalysis, 122, pp. 149-155.

[12] Folger, H.S., 1992, *Elements of Chemical Reaction Engineering*, Prentice Hall International, New York.

[13] Happel, J., Kamholz, K., Walsh, D., and Strangio, V., "Kinetics of the Isobutane-Isobutene-Hydrogen system using tracers", 1973, Industrial and Engineering Chemistry Fundamentals, 12(3), pp. 263-267.

PAPER 17

Oxidative reforming of n-butane to ethylene, propylene, and syngas on acid/base nano-hybrid catalyst

G. Tanimu [1,2], Z. S. Qureshi [1], H. S. Alasiri [1,2], S. Asaoka [1], S. Al-Khattaf [1,2]

1. *Center for Refining and Petrochemicals, King Fahd University of Petroleum & Minerals, Dhahran 31261, Saudi Arabia*

2. *Department of Chemical Engineering, King Fahd University of Petroleum & Minerals, Dhahran 31261, Saudi Arabia*

Abstract

Ni-Mx-O/support catalysts with 20 wt% Ni and 10-30 wt% Mx (= none, Bi, Mo, W, and Ga) have been studied for the oxidative reforming of n-butane to ethylene/propylene and syngas. In the catalyst, Ni oxide and Mx oxide species changed along with oxygen/n-butane feed ratio and temperature due to change in active oxygen species. The formation degree of active Ni oxide species selective towards oxidative reforming, reflected in selectivity through change in the redox and acid/base properties. The catalyst with Ni and Mo combinations shows superior performance in terms of activity and selectivity compare with other Mx loaded catalysts. Ni-Ga-O/mesoporous silica catalyst increased the ethylene/propylene-selectivity, while the Ni-Mo loading led to the high selectivity of syngas. The catalyst for oxidative conversion of n-butane showed the acid/base dependency. These characteristics agree with the catalyst performance as depicted in the catalytic scheme for selective oxidative reforming reaction.

INTRODUCTION

There exist a commercial incentive to convert low-value C_4 stream, mainly n-butane to high-value products, 1-butene, 1,3-butadiene, aromatics (benzene, toluene, and xylenes: BTX), etc in the polymer and petrochemical industries. On-purpose ethylene and propylene from n-butane is also an attractive conversion for the petrochemical industry.
In our previous work for the production of biodiesel hydrocarbon from triglyceride or fatty acid methyl ester, Ni and Mo oxide supported on gamma-Al_2O_3 (hereafter Ni-Mo-O/Al_2O_3) was exemplified as a hierarchical nano-consortium catalyst effective to the production of biodiesel hydrocarbon [1-3]. The study of the reaction mechanism revealed that CO_2 withdrawing on basic site from ester bond takes an important role in the production of short/saturated hydrocarbon (*SSHC*) and long/unsaturated hydrocarbon (*LUHC*) followed by hydrogenation.

$$\mathbf{CH_2(C_{n-1}H_{2n-1}COO)\text{-}CH(C_{n-1}H_{2n-1}COO)\text{-}CH_2(C_{n-1}H_{2n-1}COO) \rightarrow CH_3CH_2CH_3 + 3CO_2 + 3C_{n-1}H_{2n-2}}$$

Triglyceride *SSHC* *LUHC*

$$\mathbf{C_{n-1}H_{2n-3}COOCH_3 \rightarrow CH_4 + CO_2 + C_{n-1}H_{2n-2}}$$

Fatty acid methyl ester *SSHC* *LUHC*

In the follow-up work for the oxidative dehydrogenation of n-butane as on-purpose butadiene production technology, Ni or Ni/Bi oxide supported on gamma-Al_2O_3 (hereafter Ni-O or Ni-Bi-O/Al_2O_3) was exemplified as a hierarchical nano-consortium catalyst effective to butadiene production [4]. The influence of calcination, main metal modification, and support including mesoporous silica on the performance of Bi-Ni-O/gamma-Al_2O_3 catalyst for n-butane oxidative dehydrogenation to butadiene was also investigated as the improved hierarchical nano-consortium catalyst [5-8]. Direct dehydrogenation and CO_2 oxidative dehydrogenation were also investigated for producing butadiene, but they require very high temperature and resulted in lower butadiene production [9, 10]. On the other hand, the oxidative dehydrogenation of n-butane with oxygen at lower temperature has the possibility of producing ethylene and propylene as main products in the case of high conversion with less selectivity of butadiene. In this case, syngas (H_2 and CO) production was also recognized as partial oxidation product.

$$\mathbf{CH_3CH_2CH_2CH_3 + \tfrac{1}{2}O_2 \rightarrow CH_2{=}CHCH_3 + 2H_2 + CO}$$

$$\mathbf{CH_3CH_2CH_2CH_3 + O_2 \rightarrow CH_2{=}CH_2 + 3H_2 + 2CO}$$

This paper is a further follow-up to our previous works [1-10] and tends to improve the hierarchical nano-consortium catalyst in two viewpoints of metal species and support for on-purpose ethylene and propylene from n-butane through oxygenate formation and CO_2 withdrawing. The conversion concept is shown in the following scheme and the withdrawn CO_2 probably converted further to H_2 and CO within the reaction network of water-gas shift and reverse water-gas shift between H_2O/CO and H_2/CO_2.

$$\mathbf{CH_3CH_2CH_2CH_3 \rightarrow CH_2{=}CHCH_2COOH \rightarrow CH_2{=}CHCH_3 + CO_2}$$

$$\mathbf{CH_3CH_2CH_2CH_3 \rightarrow CH(COOH){=}CHCOOH \rightarrow CH_2{=}CH_2 + 2CO_2}$$

RESULTS AND DISCUSSIONS

Chemistry for reaction design of butane-to-ethylene/propylene

The typical result of reaction test is shown in **Table 1**. Based on the results in the table, a simplified reaction scheme from n-butane to ethylene/propylene (OC) including butadiene formation (BD) is shown in **Fig. 1**. The catalytic performance of the standard catalyst (Ni) and the modified catalyst (Ni-Bi) showing the main selectivity (OC and BD) with time-on-stream is shown in **Fig. 2**. The figure shows that unmodified Ni oxide without gas-phase oxygen results in high selectivity to oxygenate and cracking (ethylene/propylene) as a transition response test. Ni modification by Bi and the existence of gas-phase oxygen lead to the increased trend of butadiene production. The result means that lattice oxygen in Ni oxide behaves as active species for oxygenate formation. This fact leads to a simplified mild CPOx (Catalytic Partial Oxidation) system from n-butane to ethylene/propylene as shown in **Fig. 3**.

Table 1 Catalytic performance of none (only Al_2O_3), Ni and Ni-Bi catalysts. Catalyst: 20 wt% Ni-30 wt% Bi-O/gamma-Al_2O_3, Reaction condition: 500 °C, O_2/n-C_4H_{10} = 2.0.

Catalyst			Al_2O_3	Ni/Al_2O_3	$Ni\text{-}Bi/Al_2O_3$
n-C_4H_{10} conversion			28.1	27.7	25.1
Selectivity (n-C_4H_{10} base)	$C_2^{=}+C_3^{=}$		0.9	8.7	16.1
		$C_2^{=}$	0.9	8.2	16.1
		$C_3^{=}$	0.0	0.5	0.0
	CO		60.1	34.7	2.2
	C_1^0-C_3^0		7.1	4.9	6.2
	$C_4^{=}+C_4^{2=}$		24.7	48.1	75.4
		$C_4^{2=}$	2.0	21.9	49.7
Yield (n-C_4H_{10} base)	$C_2^{=}+C_3^{=}$		0.3	2.4	4.0
		$C_2^{=}$	0.3	2.3	4.0
		$C_3^{=}$	0.0	0.1	0.0
	CO		16.9	9.6	0.6
	H_2		7.7	20.9	2.7
	$C_4^{2=}$		0.6	6.1	12.5
	$C_3^{=}/C_2^{=}$		0.0	0.04	0.0

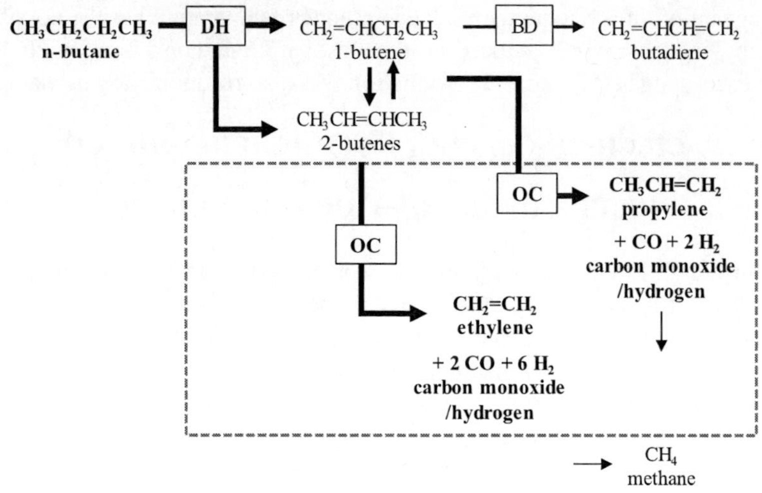

Fig. 1 Simplified reaction scheme from n-butane to ethylene/propylene including butadiene formation, DH: oxidative dehydrogenation; OC: oxygenate formation and cracking; BD: butadiene production from 1-butene.

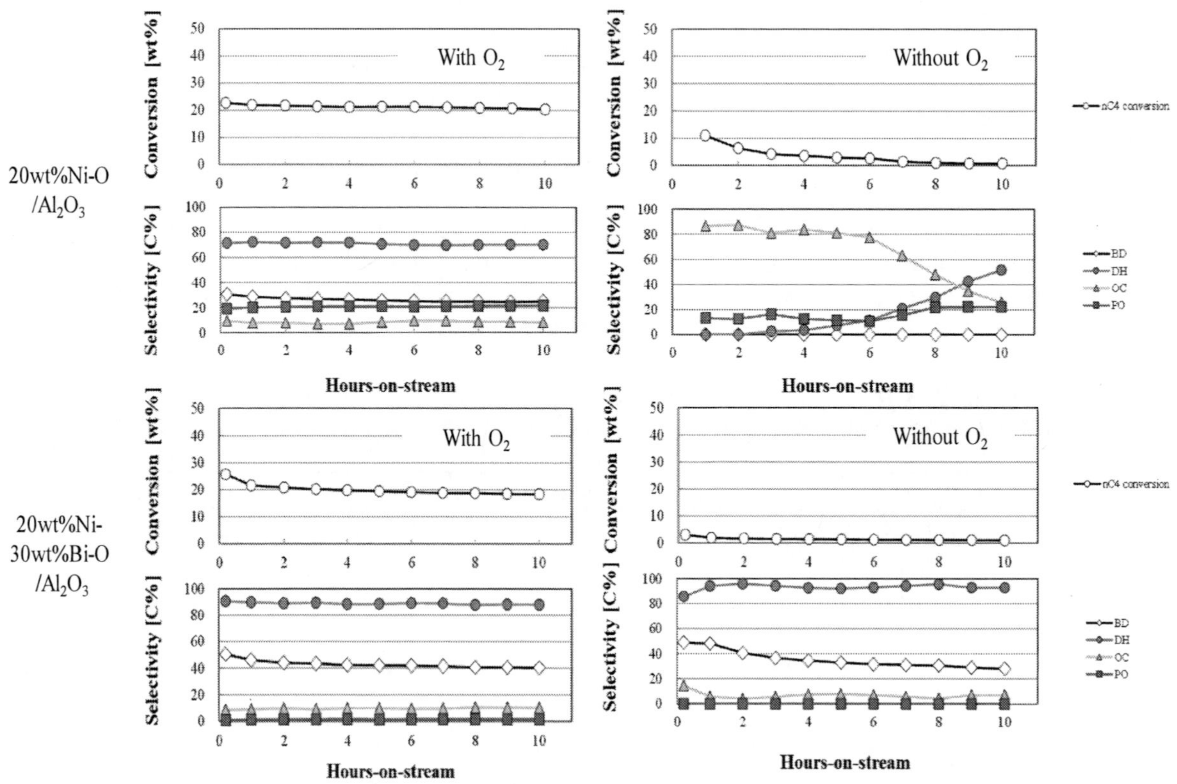

Fig. 2 Activity and reaction selectivity of n-butane oxidative dehydrogenation with/without gaseous oxygen including reactions shown in **Fig. 1**, DH: oxidative dehydrogenation; BD: butadiene production from butene; OC: oxygenate formation and cracking; PO: partial oxidation to syngas.

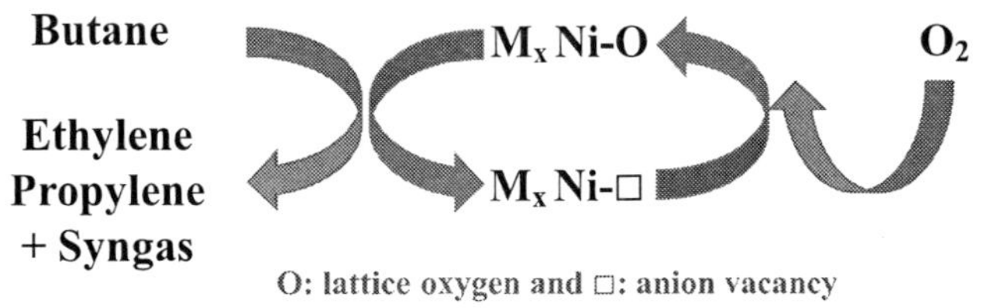

Fig. 3 Simplified mild CPOx (Catalytic Partial Oxidation) system from n-butane to ethylene/propylene.

Hierarchical nano-consortium Ni-Mx-O/Al_2O_3 catalyst

The catalyst was designed as hierarchically the active metal Ni oxide as fine nano-oxide particles at the top stabilized by cohabiting larger nano-Mx oxide particles and dispersed on fractal nano-porous alumina surface. The catalytic performance is shown in **Table 2**. The table shows that metal combination of Ni-Mo or Ni-W increased syngas production while the Ni-Bi improved olefin (ethylene/propylene and n-butenes/butadiene) formation with suppression of syngas (CO/H_2).

Hierarchical nano-hybrid Ni-Ga-O/SiO_2 foam catalyst

The catalyst was designed as hierarchically the active metal Ni oxide as fine nano-oxide particles at the top stabilized by cohabiting larger nano-Ga oxide particles and dispersed on the SiO_2 foam surface. The catalytic performance is shown in **Table 3**. The table shows that Ga oxide improved ethylene/propylene formation with suppression of butadiene production. The combination of Ni-Ga remarkably improved activity without affecting the selectivity. This fact led to the maximum yield of ethylene and propylene.

Table 2 Comparison of catalytic performance for Ni-Mx-O catalyst, Catalyst: 20 wt% Ni-30 wt% Mx-O/gamma-Al_2O_3, Reaction condition: 500 °C, O_2/n-C_4H_{10} = 2.0.

Catalyst Mx		none	Mo	W	Bi
n-C_4H_{10} conversion		27.7	57.3	23.7	25.1
Selectivity	$C_2^=+C_3^=$	8.7	3.6	8.8	16.1
(n-C_4H_{10} base)	$C_2^=$	8.2	3.0	6.1	16.1
	$C_3^=$	0.5	0.6	2.7	0.0
	CO	34.7	78.9	78.6	2.2
	C_1^0-C_3^0	4.9	0.8	6.0	6.2
	$C_4^=+C_4^{2=}$	48.1	16.7	6.6	75.4
	$C_4^{2=}$	21.9	11.8	2.8	49.7
Yield	$C_2^=+C_3^=$	2.4	2.1	2.1	4.0
(n-C_4H_{10} base)	$C_2^=$	2.3	1.7	1.5	4.0
	$C_3^=$	0.1	0.3	0.6	0.0
	CO	9.6	45.2	18.6	0.6
	H_2	20.9	(255)	8.3	2.7
	$C_4^{2=}$	6.1	6.8	0.7	12.5
	$C_3^=/C_2^=$	0.04	0.20	0.44	0.0

Table 3 Catalytic performance for Ga and Ni-Ga catalyst, Catalyst: 20 wt% Ni-30 wt% Ga-O/(SiO_2 foam), Reaction condition: 500 °C, $O_2/n\text{-}C_4H_{10} = 2.0$.

Catalyst		Ni/A	Ga/SF	Ni-Ga/SF
$n\text{-}C_4H_{10}$ conversion		27.7	17.1	35.1
Selectivity	$C_2^=+C_3^=$	8.7	37.1	27.8
($n\text{-}C_4H_{10}$	$C_2^=$	8.2	24.0	23.2
base)	$C_3^=$	0.5	13.1	4.6
	CO	34.7	28.3	41.1
	$C_1^0\text{-}C_3^0$	4.9	5.3	4.9
	$C_4^=+C_4^{2=}$	48.1	29.3	26.2
	$C_4^{2=}$	21.9	10.2	15.3
Yield	$C_2^=+C_3^=$	2.4	6.3	9.8
($n\text{-}C_4H_{10}$	$C_2^=$	2.3	4.1	8.2
base)	$C_3^=$	0.1	2.2	1.6
	CO	9.6	4.8	14.4
	H_2	20.9	9.4	7.9
	$C_4^{2=}$	6.1	1.7	5.4
	$C_3^=/C_2^=$	0.04	0.54	0.20

Base modification by addition of Mg/K

The standard catalyst was designed as hierarchically the active metal Ni oxide as fine nano-oxide particles at the top stabilized by cohabiting larger nano-Bi oxide particles and dispersed on the fractal Al_2O_3 surface. The base modification of the standard catalyst (Ni-Bi-O) was studied using Mg or K. The result is shown in **Table 4**.

Table 4 Effect of alkaline addition on catalytic performance for Ni-Bi catalyst, Catalyst: 20 wt% Ni-30 wt% Bi-O/Al_2O_3, Reaction condition: 500 °C, $O_2/n\text{-}C_4H_{10} = 2.0$.

Catalyst/Alkaline addition		none	Mg	K
$n\text{-}C_4H_{10}$ conversion		25.1	24.7	27.8
Selectivity	$C_2^=+C_3^=$	16.1	44.5	51.5
($n\text{-}C_4H_{10}$ base)	$C_2^=$	16.1	36.0	42.0
	$C_3^=$	0.0	8.5	9.5
	CO	2.2	0.0	0.0
	$C_1^0\text{-}C_3^0$	6.2	2.3	3.2
	$C_4^=+C_4^{2=}$	75.4	53.2	45.4
	$C_4^{2=}$	49.7	33.8	23.7
Yield	$C_2^=+C_3^=$	4.0	11.0	14.3
($n\text{-}C_4H_{10}$ base)	$C_2^=$	4.0	8.9	11.7
	$C_3^=$	0.0	2.1	2.6
	CO	0.6	0.0	0.0
	H_2	2.7	0.0	0.2
	$C_4^{2=}$	12.5	8.3	6.6
	$C_3^=/C_2^=$	0.0	0.24	0.22

Alkaline modification showed no effect on the catalyst activity as the n-butane conversion only slightly increased after the modification. The product selectivity, on the other hand, was affected by the modification. The unmodified Ni-Bi catalyst showed a very high selectivity to dehydrogenation products (butenes and butadiene) with low selectivity to oxygenate/cracking products. Base modification improved the cracking products (ethylene and propylene) selectivities with a continuous decreased in dehydrogenation products selectivities. K-modification is more effective in terms of the ethylene and propylene selectivities compared to Mg-modification. The modification also showed a negative effect on the partial oxidation (syngas) products selectivities.

Acid modification using siliceous MFI zeolite as support

Different species of siliceous MFI zeolites (100% silicate, Ge silicate, Si/Al=80 and 500) were used as carriers for the active metal species (Ni-Bi) to investigate whether they can be suitable as potential candidates modifying the acidity with an improved performance (activity and selectivity). The result obtained is presented in **Table 5.**

Table 5 Comparison of catalytic performance for siliceous MFI zeolite species in Ni-Bi-O catalyst. Catalyst: 20 wt% Ni-30 wt% Bi-O /zeolite support. Reaction condition: 500 °C, O_2/n-C_4H_{10} = 2.0.

Catalyst support		High Si (Si/Al=80)	High Si (Si/Al=500)	100% silicate	Ge silicate
n-C_4H_{10} conversion		10.2	10.3	16.9	37.0
Selectivity	$C_2^=+C_3^=$	39.3	33.3	20.2	29.8
(n-C_4H_{10} base)	$C_2^=$	28.6	25.1	11.5	20.7
	$C_3^=$	10.7	8.2	8.7	9.1
	CO	20.1	34.1	6.7	6.6
	C_1^0-C_3^0	2.1	5.9	0.8	5.3
	$C_4^=+C_4^{2=}$	38.4	26.6	72.4	58.2
	$C_4^{2=}$	4.0	15.1	18.7	17.1
Yield	$C_2^=+C_3^=$	4.0	3.4	3.4	11.1
(n-C_4H_{10} base)	$C_2^=$	2.9	2.6	1.9	7.7
	$C_3^=$	1.1	0.8	1.5	3.4
	CO	2.1	3.5	1.1	2.4
	H_2	0.5	0.2	0.4	6.5
	$C_4^{2=}$	0.4	1.6	3.2	6.3
	$C_3^=/C_2^=$	0.38	0.31	0.79	0.44

The activity of the catalysts decreased as the Si/Al ratio decreased. Also, the selectivities to main reaction products varied with respect to the Si/Al ratio. For 100% silicate supported catalyst, very high selectivity to dehydrogenation products was observed with low selectivity to oxygenate/cracking and syngas products. Modification of silicate with Ge decreased the selectivity to DH products with improved selectivity to oxygenate/cracking products. This catalyst gave the highest yield of ethylene and propylene. Catalysts with Si/Al =80 and 500, showed high selectivity to oxygenate/cracking products with a decreased selectivity to DH products. The acidity of zeolite supported catalysts increased with decreased in Si/Al ratio (Alumina has higher acid sites compared to silica) and acid sites favor oxygenate/cracking (ethylene and propylene) pathway.

Reaction pathway from n-butane to ethylene/propylene and syngas

The detailed reaction pathway from n-butane to the various reaction products (mainly dehydrogenation and oxygenate/cracking products) is shown in **Fig. 4.** 1,3-butadiene is obtained through two steps dehydrogenation from n-butane through butenes. Oxygenate/cracking products (ethylene and propylene) are obtained from the interactions of both n-butane and butenes (1-butene and 2-butene) with lattice oxygen species of the catalyst.

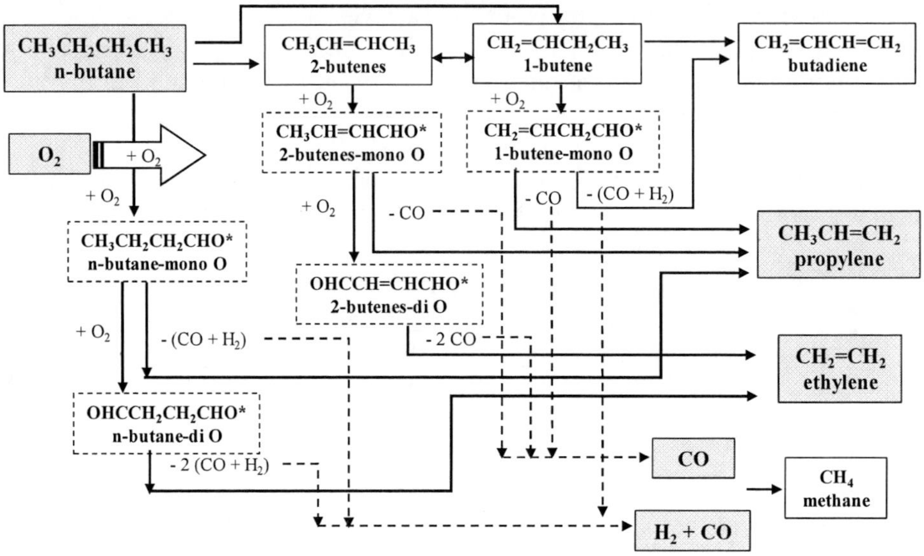

Fig. 4. Conversion route from the reactant to the products including co-product formation

Model of catalyst role for reaction from n-butane to ethylene and propylene

The role of catalyst acid/base sites and availability of lattice oxygen is shown in **Fig. 5.** The presence of moderate/strong acid sites coupled with moderate/strong base sites favors the formation of both ethylene and propylene. Intermediate 2-butene is easily formed on strong acid sites which is adsorbed on strong base sites and interacts with lattice oxygen species to form the cracking products (ethylene and propylene). On the other hand, the presence of moderate/strong base sites favors 1-butene formation which is adsorbed on weak/moderate acid site resulting in propylene formation.

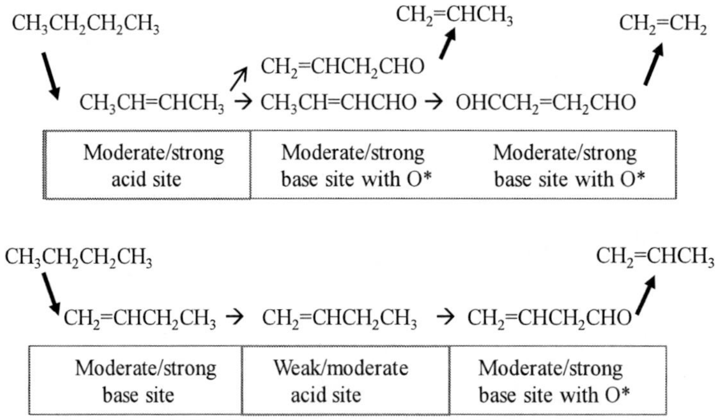

Fig. 5 Ethylene/propylene yield through moderate/strong base sites with weak/moderate acid site.

CONCLUSIONS

Ni-Mx-O/support catalysts have been studied for oxidative reforming of n-butane to ethylene/propylene and syngas, comparing from the viewpoint of loaded Mx oxide species and the combination. In the catalysts, the active oxygen species varied according to the combination of Ni and Mx oxide. The formation degree of Ni oxide species selective to oxidative reforming reflected the selectivity through redox and acid/base properties. Ni-Ga-O/mesoporous silica catalyst increased the ethylene/propylene-selectivity, while the Ni-Mo loading led to the high selectivity of syngas. The catalyst for oxidative conversion of n-butane showed the acid/base dependency. The characteristics agree with the catalyst performance in the view of the catalytic scheme for selective reaction of oxidative reforming (mild CPOx).

Acknowledgment
The authors appreciate the support of The Research Institute at King Fahd University of Petroleum and Minerals (KFUPM), and also the contribution by Japan Cooperation Center Petroleum (JCCP) for this collaborative research.

REFERENCES

[1] Kimura, T., Imai, H., Li, X., Sakashita, K., Asaoka, S., Al-Khattaf, S., *Catal. Lett.*, **143**, 1175–1181 (2013).
[2] Kimura, T., Imai, H., Li, X., Sakashita, K., Asaoka, S., Akhtar, N., Al-Khattaf, S., *Arab. J. Sci. Eng.*, **39**, 6617–6625 (2014).
[3] Imai, H., Kimura, T., Terasaka, K., Li, X., Sakashita, K., Asaoka, S., Al-Khattaf, S., *Catal. Today*, **303**, 185–190 (2018).
[4] Ajayi, B., Jermy, R., Ogunronbi, E., Abussaud, B., Al-Khattaf, S., *Catal. Today,* **204**, 189–196 (2013).
[5] Ajayi, B., Jermy, R., Abussaud, B., Al-Khattaf, S., *J. Porous Mater.,* **20**, 1257–1270 (2013).
[6] Jermy, R., Ajayi, B., Abussaud, B., Asaoka, S., Al-Khattaf, S., *J. Mol. Catal. A Chem.,* **400**, 121–131 (2015).
[7] Jermy, R., Asaoka, S., Al-Khattaf, S., *Catal. Sci. Technol.,* **5**, 4622–4635 (2015).
[8] Tanimu, G., Jermy, R., Asaoka, S., Al-Khattaf, S. *J. Ind. Eng. Chem.,* **45**, 111–120 (2017).
[9] Tanimu, G., Asaoka, S., Al-Khattaf, S., *Mol. Catal.,* **438**, 245–255 (2017).
[10] Tanimu, G., Palani, A., Asaoka, S., Al-Khattaf, S., *Catal. Today,* https://doi.org/10.1016/j.cattod.2018.06.014 (2018).

PAPER 18

Light naphtha upgrading to aromatics using metal modified MFI zeolite catalysts

Yaming Jin, Omer R. Koseoglu, and Mohammad Al-Rebh (Saudi Aramco R&DC),
Abdullah Aitani, Ziyauddin Qureshi, M. Naseem Akhtar, and Hassan Al-Asiri (KFUPM)

Light paraffinic naphtha within C5-C6 range has been historically a low-value hydrocarbon feedstock originating from refinery and gas plants. The Refining Development Team (RDT) of Refining and Upgrading R&D Division in Saudi Aramco R&D Center has been collaborating with the Center for Refining and Petrochemicals (CRP) of Research Institute of KFUPM in developing a new catalytic conversion technology that can upgrade a large quantity of surplus light naphtha in Aramco's refineries and NGL fractionation plants to value added products such as aromatics. In this paper, we will present the most recent experimental results of several catalyst systems, especially the metal modified MFI zeolite catalysts, which show superior aromatization activity and selectivity.

PAPER 19

Upgrading of unused heavy oil over iron oxide based catalysts under sub/supercritical water conditions

Takuya Yoshikawa[1)], Hisaki Kondoh[1)], Yuta Nakasaka[1)], Teruoki Tago[2)], Takao Masuda[*,1)]

[1] *Division of Applied Chemistry, Faculty of Engineering, Hokkaido University, 060-8628, Sapporo, Japan*

[2] *Department of Chemical Science and Engineering, School of Materials and Chemical Technology, Tokyo Institute of Technology, 152-8550, Tokyo, Japan*

**E-mail: takao@eng.hokduai.ac.jp*

Abstract

Upgrading of oil sand bitumen over iron oxide based catalyst was carried out under sub-/super-critical water condition. At first, the effect of reaction conditions on the product yields was investigated using both batch and flow type reactors. Reaction pressure greatly affected the product yield, revealing that the formation carbonaceous solid product was decreased with increasing pressure, and the yield of lighter component (Gas Oil and VGO component) was the highest under sub-critical water condition. Moreover, the catalytic reaction of bitumen was analyzed from the view point reaction kinetics. The decomposition reaction of VR (vacuum residual oil component) in bitumen was expressed as second order and the activation energy was calculated. The reaction pathway was analyzed using Lumping Model in detail.

Keywords

Heavy oil, Iron oxide based catalyst, Sub-/super-critical water, Kinetic analysis

1. Introduction

Petroleum has the most superior characteristics among fossil fuels as a source of energy and is still occupying a large part of energy demand in the world. However, it is said that the expansion of reserve-production ratio of crude oil decelerates in the world. Because, in addition to atmospheric residue (AR) and vacuum residue (VR) produced from conventional refinery process, there are plenty of unused carbon resources such as oil-shale, orinico-tar and oil-sand bitumen, the technological developments of upgrading these heavy oils is strongly desired.

In order to upgrade heavy oils, the ratio of hydrogen to carbon, that is H/C ratio, should be increased. Hydrogenation/hydrocracking is one of well-known processes, in which hydrogen gas under high pressure and temperature is introduced into a reactor in the existence of a catalyst, achieving desulfurization and decomposition of heavy oils [1]. However, an economic challenge remains due to the large consumptions of hydrogen gas for the hydrogenation/hydrocracking process. In addition, because hydrogen gas is generally produced from petroleum-derived feedstock such as naphtha, the employment of hydrogen for upgrading heavy oils is considered to make the supply-demand balance of hydrogen unstable. For this reason, steam catalytic cracking has attracted much attention, in which hydrogen is generated from co-supplied water. Although some aquathermolysis processes have been developed [2], water-soluble metal salts were mainly used as catalysts, making the process cost high. Accordingly, it is required to develop steam catalytic cracking process using an inexpensive solid catalysts.

Our group have desinged iron oxide based catalyst containing CeO_2, ZrO_2, Al_2O_3 (abbreviated as CeO_2-ZrO_2-Al_2O_3-FeO_X catalyst hereafter) and investigated the reaction process employing the catalyst. Fig. 1 shows the schematic of each component

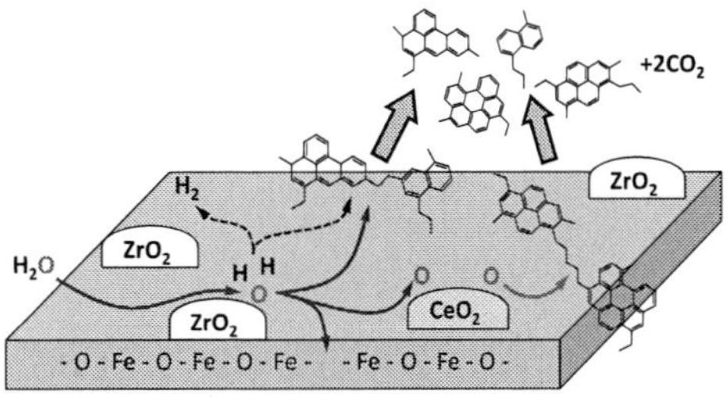

Fig. 1 Schematic of function of CeO_2-ZrO_2-Al_2O_3-FeO_X catalyst.

function consisting CeO_2-ZrO_2-Al_2O_3-FeO_X catalyst. Iron oxide is the main field of reaction, proceeding the oxidative cracking of heavy oils using the lattice oxygen. Active oxygen species from co-supplied water are generated over ZrO_2, filling the oxygen vacancy to maintain the activity and stability of catalyst. Simultaneously, active hydrogen species from water is produced to be involved with the upgrading reaction. The validity of this reaction cycle was clarified based on the detail analysis of gas and liquid products after the reactions using iron oxide based catalysts [3,4]. Because CeO_2 showed so called oxygen storage/release capacity, the property was seem to be effective for balancing the amount of active oxygen species derived from water over the catalyst. The addition of Al_2O_3 was effective for increasing specific surface area of the catalyst, and for the enhancement of physical stability of the catalyst [5].

Because approximately 80% of oil sand bitumen is discovered at great depths, steam assisted gravity drainage (SAGD) method is an important process, in which high temperature and high pressure steam is introduced into deeper layers to recover bitumen. Accordingly, the development of a catalytic reaction under high temperature and pressure water condition is expected to enable the on-site upgrading of heavy oil derived from SAGD method. It was reported that upgrading of conventional heavy oil, that is AR and VR, effectively proceeded over iron-oxide based catalyst under superheated steam condition at normal pressure [3,5,6-8]. In this study, therefore, upgrading of heavy oil, especially bitumen, under sub-/super-critical water condision was investigated over CeO_2-ZrO_2-Al_2O_3-FeO_X catalyst.

2. Experimental

Iron oxide based catalyst (CeO_2-ZrO_2-Al_2O_3-FeO_X) was prepared by a co-precipitation method, using mixed aqueous solution of $Ce(NO_3)_3{\cdot}6H_2O$, $ZrO(NO_3)_2$, $Al(NO_3)_3{\cdot}9H_2O$, and $Fe(NO_3)_3{\cdot}9H_2O$ as precursors. The composition of CeO_2, ZrO_2, and Al_2O_3 were set as 2.5, 7.5 and 7.0 wt%, respectively. The crystal structure and porous property of the catalyst were analyzed by X-ray diffraction and N_2 adsorption isotherms, respectively.

Upgrading of bitumen over CeO_2-ZrO_2-Al_2O_3-FeO_X was carried out using batch and flow type reactors as shown in Fig. 2. Canadian oil sand bitumen diluted with benzene to 10 wt% for reducing its viscosity was employed as a feedstock. The detail reaction conditions were described in the section of results and discussion.

After the reaction, the gas and liquid products were separately collected. Gas products were analyzed by two GCs equipped with activated charcoal and Porapak Q columns using TCD and FID detectors, respectively. After separating the liquid product into water and oil phases with centrifuge, the resulting oil phase was analyzed by HPLC with a PL-gel Mixed-D column using RI detector. The molecular weight distribution of bitumen and oil product acquired from HPLC was converted into carbon number distributions by assuming the unit structure of heavy oil to be -CH_2-. Then, the liquid product yield was divided into the following three groups according to the carbon number: gas oil (Gas Oil, carbon numbers of 14–20), vacuum gas oil (VGO, carbon numbers of 21–40) and vacuum residual oil (VR, carbon numbers over 41). Carbonaceous solid deposited on the reactor wall was designated as residue, and coke deposited on the catalyst was analyzed by an elemental analyzer. The above experimental procedure was described in detail in our previous study [4, 9].

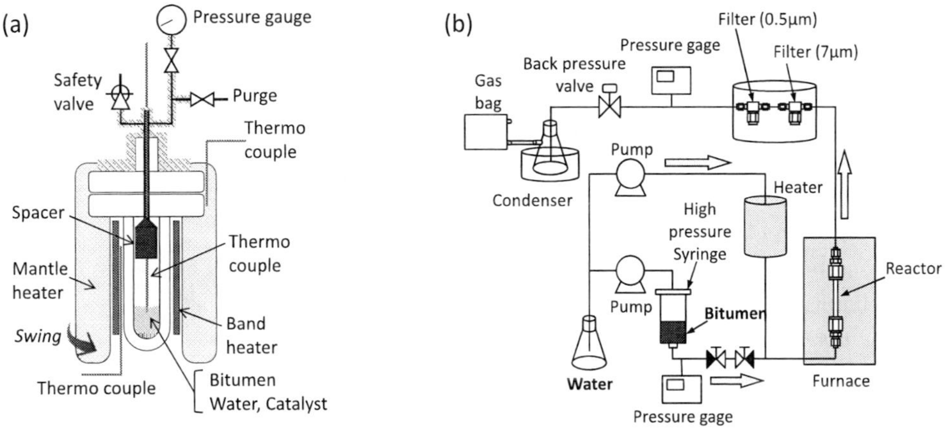

Fig. 2 Reaction experimental apparatus
(a) Batch type reactor, (b) High pressure fixed bed flow type reactor

3. Results and discussion

3.1. Effect of reaction conditions using batch type reactor

At first, the effect of reaction conditions on the product yields and the catalytic stability was investigated in the reactions of bitumen over iron oxide based catalyst using batch type reactor [9]. Fig.3 shows the effect of reaction pressure on the yields of lighter components (Gas Oil and VGO) and coke after upgrading of bitumen at 693 K for 1 hr. High yield of lighter components is obtained under high pressure (8.6–22.3 MPa), reaching approx. 70 C-mol%, and the formation of coke were decreased with increasing reaction pressure. We have reported that catalytic reactions under sub-/super-critical solvent condition were effective for suppressing the coke formation on catalysts in the reactions such as decomposition of lignin-derived liquid over iron oxide based catalyst [10], and alkylation and transalkylation reactions over zeolite catalysts [11,12]. Coke precursor during the reaction of bitumen was considered to be dissolved or dispersed in water under sub-/super-critical water condition, resulting in the suppression of coke formation. On the other hand, the recombination reactions of the lighter components seemed to be proceeded under super critical water condition, leading to the decrease in the yield of light component.

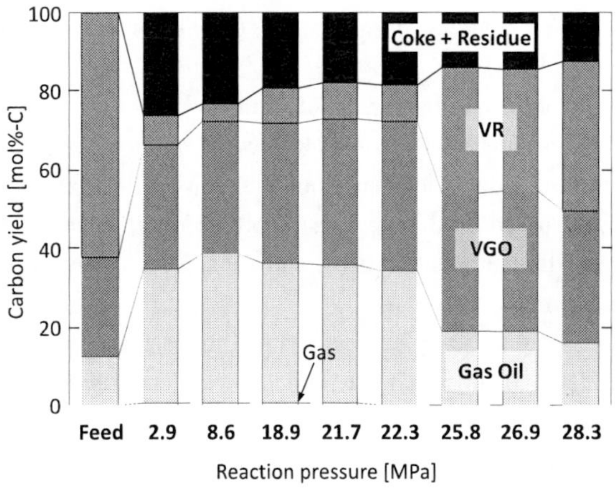

Fig. 3 Effect of reaction pressure on product yield for reaction of bitumen over CeO_2-ZrO_2-Al_2O_3-FeO_X using a batch reactor.Reaction conditions: 10 wt% bitumen solution in benzene, $W_{cat}/W_{Bitumen} = 4$, $W_{H2O}/W_{Bitumen} = 20$, $P = 2.9$–28.3 MPa, $T = 693$ K, t (reaction time) = 1 hr

3.2. Continuous upgrading of bitumen using high pressure flow type reactor

Based on the results using batch type reactor in the previous section 3.1, a high pressure flow type reactor was applied to continuous upgrading of bitumen over iron oxide based catalyst. Fig. 4 shows the effect of reaction atmosphere on the upgrading of bitumen (superheated steam and sub-/super-critical water conditions). The yield of lighter components increased to c.a. 70 C-mol%, and the coke amount decreased to 6 C-mol% under sub-critical water condition, which was consistent with the result obtained using batch type reactor shown in Fig. 3.

Durability of catalytic reaction was also examined. The upgrading of bitumen was carried out at 693 K under sub-critical water condition (19 MPa) for 6 hr. The similar product yield to the reaction for 2 hr was obtained and the catalyst structure kept the original hematite after the reaction [9].

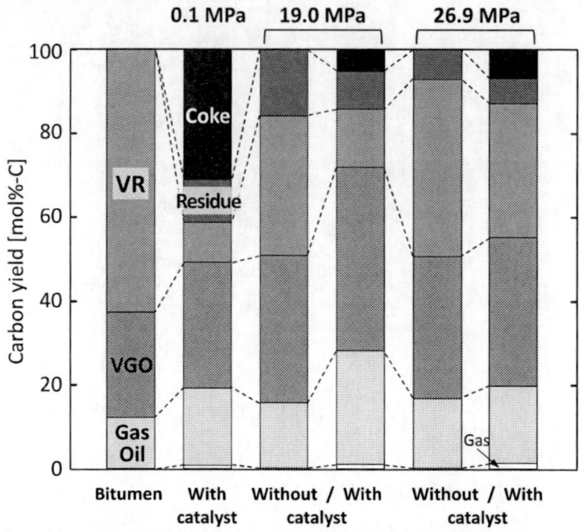

Fig. 4 Effect of reaction pressure on product yield for reaction of bitumen over CeO_2-ZrO_2-Al_2O_3-FeO_X using a flow type reactor. Reaction conditions: 10 wt% bitumen solution in benzene, $W_{cat}/F_{Bitumen} = 4$ h, $F_{H2O}/F_{Bitumen} = 20$, $P = 0.1$–26.9 MPa, $T = 693$ K, t (reaction time) = 2 hr.

3.3. Kinetic analysis for upgrading bitumen over iron oxide based catalyst

Continuous upgrading of bitumen using iron oxide based was investigated further from the view point of reaction kinetics. Because VR occupied a large part of bitumen, VR was regarded as reactant in the reaction of bitumen, and the decomposition rate of VR was analyzed. We adopted second order reaction for the decomposition of VR based on the previous reports [13,14], calculating the reaction rate constant by integral method. Arrhenius plot exhibited good linearity, and the apparent activation energy E_a for the decomposition of VR was calculated as 132 kJ/mol.

Because heavy oil contains multimolecular, the reaction rate analysis was carried out in detail using Lumping Model [15], in which the products were classified into the following lumps: VR, VGO, Gas oil and Gas, Coke. Reaction network of each lump was built up, and the reaction rate constant of each route was calculated. Fig. 5 shows the change of product distribution with respect to W/F at 693 K and 19 MPa, in which plot and solid lines express experimental data and analytical values, respectively. Fig. 5 revealed that analytical values were in good agreement with experimental data.

Then, the production/consumption route of each lump was examined based on the obtained rate constants. Fig.6 shows the outline of analytical results using Lumping Model. The bold arrows in Fig.8 represented the main reaction route. In short, it was revealed that VR, heavy component in bitumen, was consecutively converted into VGO, Gas Oil and Gas (VR ⇒ VGO ⇒ Gas Oil ⇒ Gas). Moreover, comparing the pathway from VR to VGO with to Coke, VR was preferentially converted into VGO. That is to say, kinetic analysis showed that the formation of coke was effectively suppressed in the reaction of bitumen under sub-critical water condition using iron oxide based catalyst.

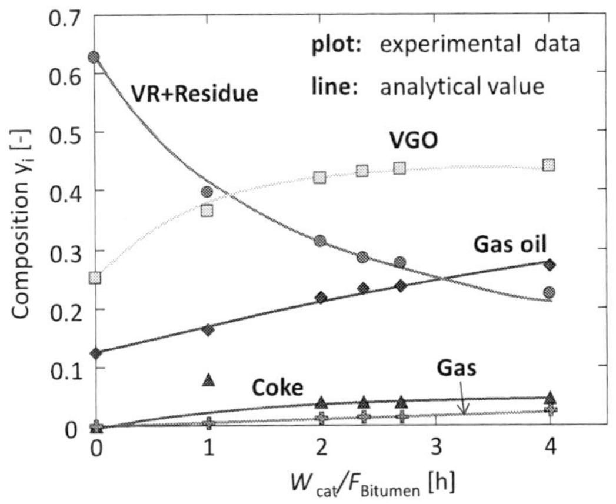

Fig. 5 Variation of product distribution with respect to $W_{cat}/F_{Bitumen}$.
Reaction conditions: 10 wt% bitumen solution in benzene, $W_{cat}/F_{Bitumen}$ = 0–4 h, $F_{H2O}/F_{Bitumen}$ = 20, P = 19 MPa, T = 693 K, t (reaction time) = 2 h.

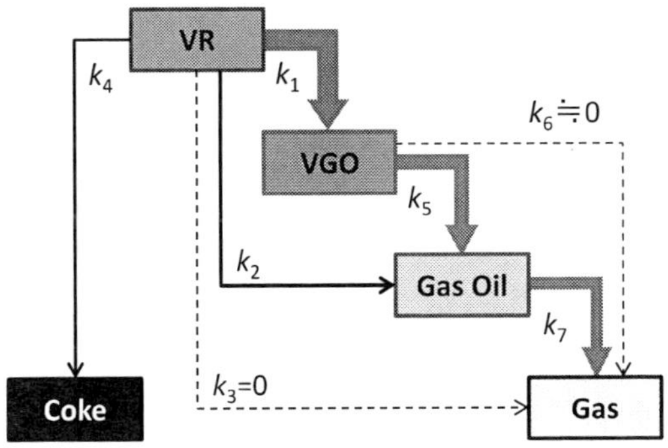

Fig. 6 Outline of reaction network obtained from the analysis using Lumping Model.

4. Conclusion

Upgrading of bitumen using iron oxide based catalyst was carried out under sub-/super-critical water condition, and the reaction process was investigated from the view point of reaction engineering. The upgrading reaction under sub-critical water condition achieved c.a. 70 C-mol% of light component (Gas Oil and VGO) with 6 C-mol% of coke at 693 K and 19 MPa using a high pressure flow type reactor. Reaction kinetic analysis was investigated in detail using Lumping Model, validating the effectiveness of catalytic reactions under sub-critical water condition. Accordingly, as shown in Fig. 7, the reaction process using iron oxide based catalyst is expected to be applied for the on-site upgrading of heavy oil including bitumen derived from SAGD method.

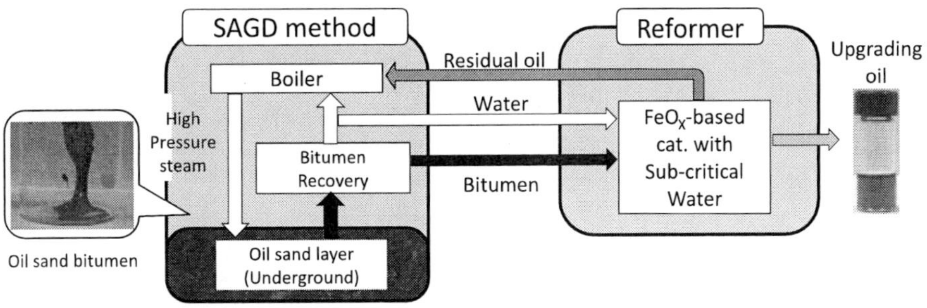

Fig. 7 Proposal of process for on-site upgrading of bitumen derived from SAGD method using iron-based oxide catalyst.

Acknowledgement

This work was supported by innovative next generation petroleum refining technology development program from the Japan Petroleum Energy Center (JPEC) from 2007 to 2011.

Nomenclature

W_{cat}	: catalyst amount loaded in a batch reactor	[kg]
$W_{Bitumen}$	: bitumen amount loaded in a batch reactor	[kg]
W_{H2O}	: water amount loaded in a batch reactor	[kg]
$F_{Bitumen}$	: flow rate of bitumen using a flow type reactor	[kg h^{-1}]
F_{H2O}	: flow rate of water using a flow type reactor	[kg h^{-1}]

References

1. R. Sahu, B.J. Song, J.S. Im, Y. Jeon, C.W. Lee, *Ind. Eng. Chem.*, **27**, 12-24 (2015).
2. O. Muraza, A. Galadima, *Fuel*, **157**, 219-231 (2015).
3. E. Fumoto, Y. Sugimoto, S. Sato, T. Takanohashi, *J.Jpn. Petrol. Inst.*, **58**, 329-335 (2015).
4. H. Kondoh, N. Hasegawa, T. Yoshikawa, Y. Nakasaka, T. Tago, T. Masuda, *Energy Fuels*, **30**, 10358-10364 (2016).
5. S. Funai, E. Fumoto, T. Tago, T. Masuda, *Chem. Eng. Sci.*, **65**, 60-65 (2010).
6. E. Fumoto, T. Tago, T. Tsuji, T. Masuda, *Energy Fuels*, **18**, 1770-1774 (2004).
7. E. Fumoto, T. Tago, T. Masuda, *Energy Fuels*, **20**, 1-6 (2006).
8. S. Funai, T. Tago, T. Masuda, *J. Jpn. Inst. Energy*, **88**, 187-192 (2009).
9. H. Kondoh, Y. Nakasaka, T. Kitaguchi, T. Yoshikawa, T. Tago, T. Masuda, *Fuel Process. Technol.*, **145**, 96-101 (2016).
10. T. Yoshikawa, S. Shinohara, T. Yagi, N. Ryumon, Y. Nakasaka, T. Tago, T. Masuda, *Appl. Catal. B-Environ.*, **146**, 289-297 (2014).
11. G. Watanabe, Y. Nakasaka, T. Masuda, *J.Jpn. Petrol. Inst.*, **60**, 146-153 (2017)
12. T. Yoshikawa, T. Umezawa, Y. Nakasaka, T. Masuda, *Catal. Today*, *in press.*
13. A. Del Bianco, N. Panariti, S. Di Carlo, P.L. Beltrame, P. Carniti, *Energy Fuels*, **8**, 593-597 (1994)
14. E. Fumoto, A. Matsumura, S. Sato, T. Takanohashi, *Energy Fuels*, **23**, 5308-5311 (2009).
15. T. Masuda, *J.Jpn. Petrol. Inst.*, *submitted.*

PAPER 20

Zeolite catalyst development for enhancing *p*-xylene selectivity

Mohammed Z. Albahar[1], Chaozhou Li[2], Vladimir L. Zholobenko[3], Arthur A. Garforth[2]
[1] Research and development Centre, Saudi Aramco, Dhahran, KSA
[2] School of Chemical Engineering and Analytical science, The University of Manchester, UK
[3] School of Chemical and Physical Sciences, Keele University, UK

Xylenes are highly demanded materials used as intermediates in the petrochemical industry. However, p-xylene is the most desired isomer amongst the xylenes and it has been increasing in recent years with an annual growth rate of 6-8%. This study demonstrated the effect of crystal size and the modification by silica deposition on the surface of the zeolite. MFI zeolites with various crystal sizes (5, 50 and 100 µm) were synthesized and compared with the commercially obtained having a crystal size of 0.5 µm. The increase in crystal size enhanced p-xylene selectivity which was attributed to the diffusion constraints imposed by the longer diffusion path lengths of large crystals. The highest p-xylene selectivity (58 %) was achieved over the catalyst with the largest crystal size 100 µm. Different post modification methods were applied in this study in attempt to suppress the fast isomerization reaction by controlling the external acid sites. The disproportionation reaction over the modified catalyst resulted in significantly improved p-xylene selectivity at 84 %, especially over silica modified large crystals 5 µm. The study emphasized the importance of zeolite structure, morphology and acidity in maximizing p-xylene selectivity.

PAPER 21

MgO and/or $Mg(OH)_2$ -blended Poly Ethylene Glycol Composites as a Shape-Stabilized Phase Change Material for Solar Thermal Energy Storage

Md. Hasan Zahir[1]*, Sultan Allihyani[2], Mohammad M. Hossain[2], Fahad A. Al-Sulaiman[1,3]

[1]Center of Research Excellence in Renewable Energy (CoRERE)-Research Institute,
[2]Department of Chemical Engineering,
[3]Department of Mechanical Engineering, Fahd University of Petroleum & Minerals, Dhahran, Saudi Arabia

Abstract

MgO and $Mg(OH)_2$ with porous structure were prepared using different types of precipitating agents, including NaOH, KOH, NH_3 with paomic acid (PA) and NH_3CO_3 for the first time by hydrothermal method at 200 °C for 24 h. Among the used agents, only NaOH provided single phase MgO powder, after calcination of the precipitate for 2h at 400°C. The NH_3 with paomic acid system increased the thermal stability of the as-synthesized $Mg(OH)_2$. The synthesized MgO and/or $Mg(OH)_2$ were used to confine blended Poly Ethylene Glycol (PEG, MW 1000 and 6000, 1:1 %wt) and form a phase change composite where MgO and $Mg(OH)_2$ served as a matrix to accommodate blended PEG. Both MgO and $Mg(OH)_2$ matrix protected the PEG liquid leakage during the phase transition. The SEM analysis of the composite showed that considerable amount of PEG was occupying within the pores, well-dispersed on the surfaces of the magnesium oxide porous structure which stabilized by capillary forces and the surface tension. The results of the differential scanning calorimetry indicated that the transition freezing temperature of the composite was 28.50 °C, while the melting point was 58.10 °C in the case of $Mg(OH)_2$ matrix. The latent heat of melting was found to be 165 J/g, which is significantly higher than those with the commonly available materials. This property indicates that the synthesized MgO and $Mg(OH)_2$ composite is a promising material for solar applications including, storing energy with good thermal stability.

---

*Corresponding author; E-mail address: hzahir@kfupm.edu.sa (Zahir)

1. Introduction

Attention on phase change materials (PCMs) has increased significantly as a medium for solar energy storing. These materials are capable of absorbing and releasing huge quantities of latent heat during phase transformation [1,2]. Organic PEG based PCM has gained growing attention because of their no supercooling and phase transfer properties. It is categorized as a good PCM due to its high chemical stability, high latent heat and eventually due to its suitable melting point. The melting point of PEG based PCM ranges from 3.2 to 68.7, which is mainly determined by the molecular weight of the polymer [3,4]. The low thermal conductivity and furthermore leakage of the molten phase of these materials requires further improvement in order to apply PEG based PCM in larger scale applications. In this regard, a support or matrix can play an important role preventing the leakage and enhancing thermal conductivity. Depending on the properties of surface tension and capillary forces, the organic PCMs can be mixed/bound inside the inorganic matrixes pores. This approach allows PCMs to store/even release heat in the melting/solidifying ranges without any leakage of the molten material. Generally, PEG of molecular weight 6000 and 1000, is used as a PCM for the polymer in certain desired range of melting and freezing temperatures. The idea of using the Paomic acid (PA) is to prevent the reaction of CO_2 to form Magnesium Carbonate $MgCO_3$. Presumably, PA acts as a reducing agent and/or stabilizer. PA forms a strong complex of $Mg(OH)_2$ due to the wide availability of OH^- ions in its structure. In this study, MgO and $Mg(OH)_2$ were prepared by hydrothermal method at low temperature in presence or absence of PA. Then the synthesized phase pure MgO or $Mg(OH)_2$ was mixed with blended PEG (MW 1000 + 6000 with 1:1 %w.t) for the first time for fabricating a noble PCM.

2. Methodology

2.1. Materials

PEG polymers with an average molecular weight of 6000 and 1000 were purchased from Guangzhou Chemical Agent Company (Guangzhou, China). Magnesium nitrate hexahydrate (Mg(NO3)2·6H2O) and absolute ethanol were purchased from Shanghai Sinopharm Chemical Reagent Co., Ltd. Paomic acid and ammonia were purchased from Sigma-Aldrich.

2.2. Hydrothermal Method

6.00 g of Mg(NO3)2·6H2O was dissolved in distilled water. 0.20 g paomic acid was dissolved also in distilled water and added into the above solution. NH3 (4.1 mol cm–3) was added to the solution for co-precipitation of the metallic ions. pH of the solution was adjusted 9.00. The suspended precipitate was transferred to a plastic container. The container was put into a steel vessel. With the tight closing of the container mouth, the steel container was put in a furnace where the hydrothermal reaction was operating at 200°C in 24 hours. The powder product was firstly washed with distilled water and ethanol, dried at 120°C in an oven [5].

2.3. Preparation of the PCM composite

The PCM composite comprising PEG/MgO or Mg(OH)2cwas prepared by dissolving PEG(6000-1000) (0.5 g) in absolute ethanol (100 mL), and the mesoporous Mg(OH)2 or MgO (0.2 g) was then added to the solution. After stirring for 4 h, ethanol was evaporated at 80°C for 24 h. The PEG/MgO or Mg(OH)2 composite PCM was collected for further characterization.

2.4. Characterization

The X-ray diffraction (XRD) patterns of the samples were carried out using an X-ray diffractometer, operated at 40 kV and 30 mA. The morphology and microstructures of the MgO material and PEG/ MgO composite were observed using a scanning electron microscope (SEM). The Fourier transform infrared (FT-IR) spectra were recorded in which the operating condition was from 400 to 4000 cm–1. The phase change temperature and latent heat of the samples were measured using a differential scanning calorimeter (DSC). DSC measurements were conducted by heating 10 mg samples sealed in an aluminum pan at a heating rate of 5°C/min under a constant stream of argon at a flow rate of 20 mL/min.

3. Results and discussion

3.1. Microstructures of mesoporous MgO or Mg(OH)2 matrix

The effect of four types precipitating agents namely, NaOH, KOH, NH_3 with paomic acid and NH_3CO_3 were tested for the preparation of MgO from its nitrate salt. Only $Mg(OH)_2$ was formed after addition NH_3 with paomic acid and their XRD patterns of $Mg(OH)_2$ were unchanged even after calcination at 400°C for 2h. In the case of KOH, the as-synthesized product were a mixture of $Mg(OH)_2$ and MgO. $MgCO_3$ was formed by treating with NH_3CO_3. In all cases, the desired product (i.e. the porous support of MgO) was not obtained. However, the as-synthesized powders had $Mg(OH)_2$ when NaOH was used and phase pure MgO was obtained just after calcination 400°C for 2h [Fig. 1, right, top].

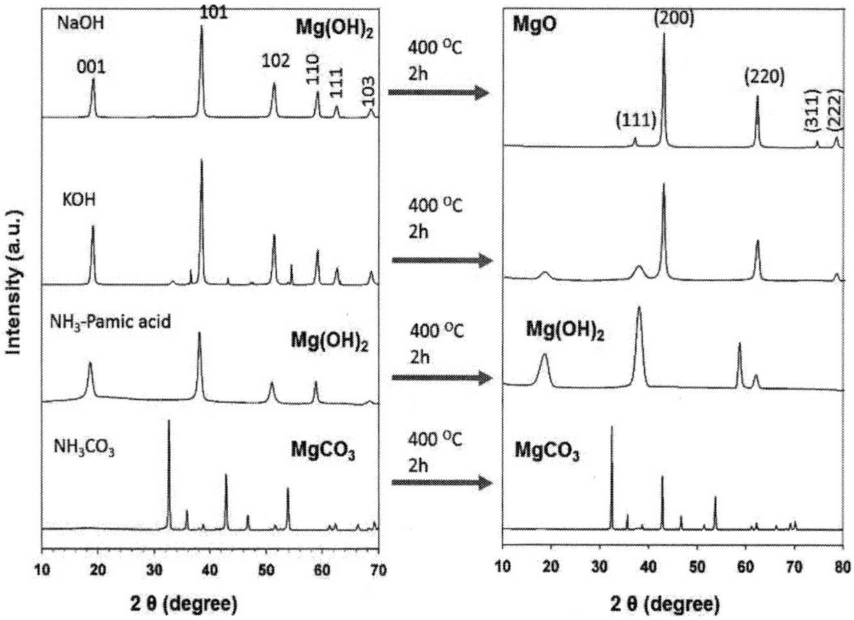

Fig. 1. Left side: XRD patterns of as-synthesized powders with precipitating agents NaOH, KOH, NH_3 with paomic acid and NH_3CO_3 [from top], right side: phase pure MgO, mixed phases, $Mg(OH)_2$ and $MgCO_3$ after calcination the as-synthesized powders at 400 °C for 2h.

All the diffraction peaks in the pattern of the powder produced correspond to the cubic MgO (JCPDS 45-0946) peaks with space group Fm-3m (225). The strong and clear peaks reveal good crystallinity of the as-obtained product. No characteristic peaks for other impurities were observed. The cell parameter calculated from the XRD pattern is a = 4.215 which is consistent with that quoted in JCPDS 45-0946 (a = 4.2112). Therefore, we conclude that the powder obtained is cubic MgO [Fig. 1, right, top]. To the best of our knowledge, this is the first repot of MgO preparation using NaOH after calcination at low temperature. It seems hydrothermal method had good impact over as-synthesized powders, which was obtained by NaOH. In the case of NH_3 mixed with paomic acid, PA probably increased the thermal stability of $Mg(OH)_2$. XRD patterns of $Mg(OH)_2$ are in good agreement with the JCPDS card number of 7-239. The significant peak broadening indicates the $Mg(OH)_2$ has a very small grain size. Figure 2 show the X-ray patterns of PEG (MW 1000 +6000) and $Mg(OH)_2$/PEG composite. The composite components $Mg(OH)_2$/PEG were well-mixed because the presence $Mg(OH)_2$ as well as PEG phases as shown in $Mg(OH)_2$-PEG. The interactions between PEG and $Mg(OH)_2$ were also characterized by using FTIR spectroscopy at room temperature. The FT-IR spectra shows that PEG/ $Mg(OH)_2$ composite was well mixed [4]. SEM image apparently shows the magnesium oxide had very small nanosize particles with porous structure. The distribution of elements in SEM mapping indicates that all elements are homogeneously distributed.

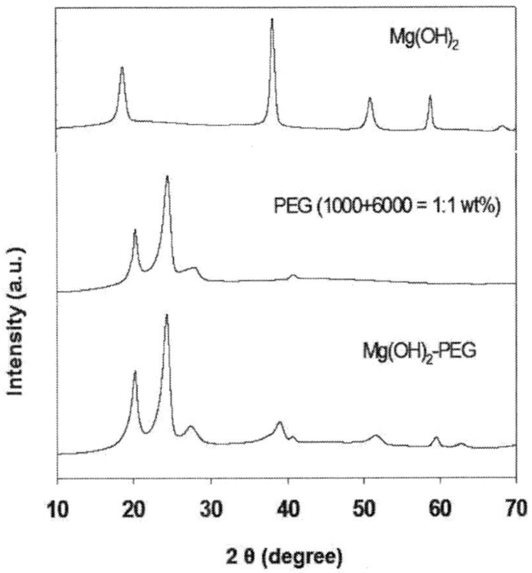

Fig. 2: (a) XRD patterns of $Mg(OH)_2$, (b) PEG(1000-6000) (c) PEG/$Mg(OH)_2$ (PA-added) composites

3.2. Thermal properties of the PEG/ Mg(OH)2 composite PCM

Several groups have attempted to address the PEG leakage problem by encapsulating PEG; however, encapsulation alone created another problem in the form of supercooling. Moreover, PCM encapsulation is a complicated and laborious process. By contrast, shape stabilization is advantageous for the preparation of PCMs. The shape stabilization process provides no or little supercooling, and no cells are formed. Figure 3 show the melting–freezing curves of PEG-6000 and PEG 1000- 6000 mixed with $Mg(OH)_2$ samples. The pure PEG-6000 had melting point of 55.30 °C and freezing point of 28.50 ° (Fig. 3a). The latent heat value was 165 J/g. A value lower than that of pure PEG, probably due to the presence of a certain weight fraction of the porous MgO matrix in the composite The microstructures and the phase change properties of the composite remained unchanged after undergoing the thermal cycles. Thermal cycling tests revealed that the PEG/MgO provided excellent thermal reliability.

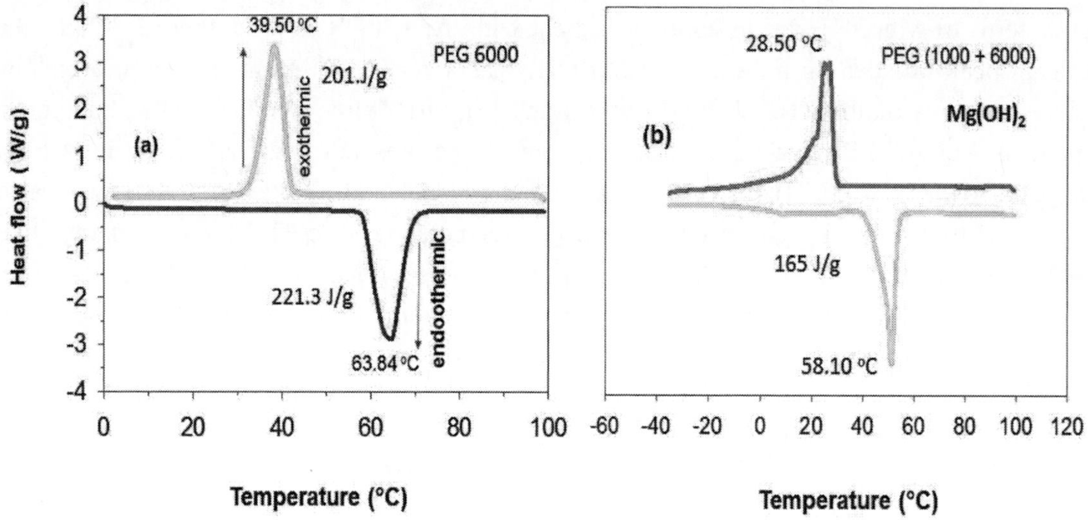

Fig 3: Melting–freezing DSC curves of (a) pure PEG-600 and (b) PEG (4000+6000) /$Mg(OH)_2$ composite.

4. Conclusions

An unique MgO and $Mg(OH)_2$ preparation method was developed using hydrothermally method. The presence paomic acid increased the thermal stability of $Mg(OH)_2$. A phase change composite material was prepared by impregnation. $Mg(OH)_2$ and MgO was used as the support material and PEG (1000-6000) was used as the phase change material. The leakage issue of PEG melting was resolved. The latent heat value and thermal properties are very promising. The newly developed PCM are viable for real application in hot climate particularly for gulf region.

Acknowledgement: The authors would like to acknowledge the support provided by the Deanship of Scientific Research (DSR) at King Fahd University of Petroleum & Minerals (KFUPM) by funding this work through project No. IN-171036.

References

1. S.A. Mohamed, F.A. Al-Sulaiman, N.I. Ibrahim, M.H. Zahir, et al. Renew. Sustain. Energy Rev. **70** (2017) 1072-1089.
2. Zalba, J.M. Marín, L.F. Cabeza, H. Mehling, Appl. Therm. Eng. **23** (2003) 251-283.
3. K. Pielichowska, K. Pielichowski, Prog. Mater. Sci. **65** (2014) 67–123.
4. A. Najjar, A. Hasan, Energy Convers. Manage. **49** (2008) 3338-3342.
5. Y. Hao, Xiankun Shao, Tongxuan Liu, Benxia Li, Shibin Nie, **604** (2015) 45-51.
6. Md. Abdul Aziz, Md. Hasan Zahir, M. Nasiruzzaman Shaikh, **28** (2017) 3226-3233.

PAPER 22

Nuclear magnetic resonance (NMR) spectroscopy at Saudi Aramco's R&D Center: where synergy is an opportunity

Qasim Saleem
Saudi Aramco R&DC
Dhahran

Nuclear Magnetic Resonance (NMR) spectroscopy provides atomic-level insight of liquids and solid materials. The NMR lab at the Technical Services Division of the Saudi Aramco R&D Center is equipped with two high-field NMR magnets capable of multinuclear measurements, as well as a low-field benchtop NMR spectrometer for rapid sample measurements. The lab is staffed by 3 dedicated personnel, who are mandated to collaborate and provide support to the field, R&DC groups and external partners. Along with the routine characterization of a variety of hydrocarbon streams, the team has also contributed extensively to complex polymer characterizations to understand structure-function relationships for optimal gas separation. Recently, the team has expanded the lab's capabilities to include solid-state characterization of zeolitic catalysts, which has attracted proponents from Saudi Aramco R&D groups as well as external collaborators, such as the Center of Research Excellence in Nanotechnology (CENT) at KFUPM. In this presentation, the synergy between the team, equipment and competency will be showcased to show the lab's mandate to match, step-for-step, increasing research aspirations.

PAPER 23

Catalytic Conversion of Saudi Alpha-olefins to Energy-saving Value-added Polymeric Flow Aids

M. Atiqullah, CRP, RI-KFUPM

Talk Abstract

Saudi Arabia is a leading crude oil producing and refining country. The transportation of crude oil and refinery products consumes a lot of highly expensive energy. Hence, the energy must be conserved, especially without changing the pipeline infrastructure. Currently, Saudi Aramco (hence Saudi Arabia) imports the highly expensive energy-saving flow aids, called drag reducing polymers (DRPs). A typical DRP is at least 10 times costlier than the relevant *alpha*-olefin(s). Therefore, the efficient transport of crude oils and refined products is a challenge, and producing DRP locally is a subject of strategic importance. Market research reports already predict steady growth of DRP business worldwide, including Saudi Arabia.

The KFUPM Center for Refining & Petrochemicals (CRP) recently completed a multimillion SR project in this area for Saudi Aramco. Through this project, a catalyst system has been developed that converts SABIC linear alpha-olefins to DRPs with much higher activity. The novelty of this catalyst system has already been protected through a US patent. This proposed talk will present the concept underlying this novel catalyst system, its comparative performance, niche for value creation, challenges facing high-performance polymerization process and product development, and business prospect.

28th Annual Saudi-Japanese Symposium
12–13 November, 2018

Catalytic Conversion of Saudi Alpha-olefins to Energy-saving Value-added Polymeric Flow Aids

Atiqullah M[1*], Al-Sarkhi AS[2], Al-Thenayan FM[3], Al-Malki AR[3], Xu W[3], Hossaen A[1]

[1]Center for Refining & Petrochemicals, Research Institute, King Fahd University of Petroleum & Minerals, Dhahran 31261, Saudi Arabia
[2]Department of Mechanical Engineering, King Fahd University of Petroleum & Minerals, Dhahran 31261, Saudi Arabia
[3]Saudi Aramco R&D Center, Dhahran 31311, Saudi Arabia

*Corresponding author; e-mail: matiq@kfupm.edu.sa.

ABSTRACT

This paper originates from a 3.2 million SR project completed for Saudi Aramco. This research adds a new dimension titled *Application and Synthesis of Amorphous Polyolefins* to the polymer research of the Center for Refining & Petrochemicals at KFUPM Research Institute. It summarizes the background as well as the related technical, business, and economic prospect, which align with Vision 2030. However, the major emphasis is on the novel catalyst system that has been developed to convert SABIC α-olefins into energy-saving drag reducing polymers (DRPs) with higher activity. Saudi Aramco and KFUPM have jointly patented this catalyst system, synthesized using a new concept that applies Lewis bases featuring simultaneously steric hindrance and electronic effect. The present catalyst prepares DRP flow aid(s) at room temperature without applying a cryogenic system. The resulting product almost triples the rate of transportation of a selected grade of refinery product with a saving of about 50% pumping energy at ppm level pipeline concentration. Also, this very easily dissolves in the above refinery product. Because of these advantages, installation of new pumps or modification of existing pipeline will be hardly necessary. This will save additional infrastructure cost. The technical challenges broadly facing this research area have been particularly discussed. Future research niches have been outlined. Our vision to develop an interdisciplinary world-class DRP research program at KFUPM has been articulated.

1. Introduction

Saudi Arabia is a leading crude oil producing and refining country in the world. See Figure 1. When crude oil and refinery products are transported through a pipeline, the pressure drastically drops due to interfacial friction between the pipe wall and the flowing fluid. Consequently, the desired throughput cannot be maintained. Increasing the input pressure and introducing higher flow rates seem to be quick and facile solutions. However, design limitations on pipelines limit the level of pressure that can be applied. The pressure further drops when fluids are transported over long distances. The consequences are increased equipment and operational costs. Therefore, the efficient transport of crude oils and refinery products is a challenge [Wilkinson et al. 1988; Al-Roomi et al. 2004; Mowla and Naderi 2006; Hasan et al. 2010; Martinez-Palou et al. 2011; and Al-Surakji et al., 2017]. To overcome the above undesirable pressure drop occurring in turbulent flow pipeline transport, ultrahigh molecular weight (UHMW, MW $\geq$ a million Dalton) polymeric drag reducing polymers (DRPs) are commonly used.

Consequently, the fluid flow rate increases at a given pumping pressure. It is Toms who first published DRP research in 1949 [Toms 1949].

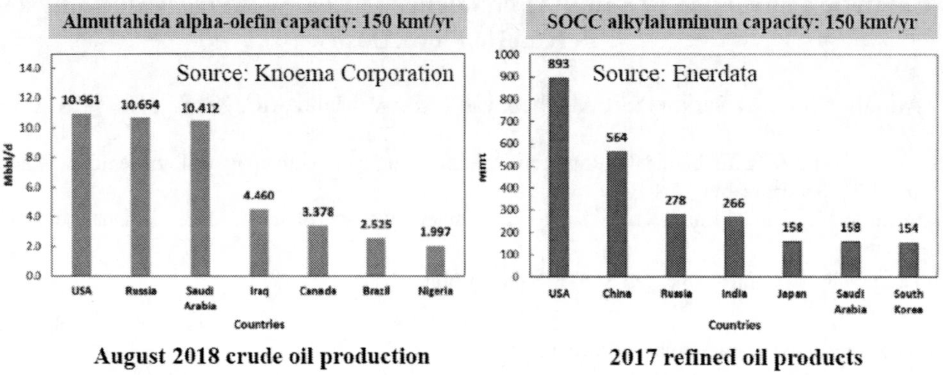

Figure 1. Production of August 2018 crude oil and 2017 refined oil products of selected countries.

As stated above, the transportation of crude oil and refinery products consumes a lot of highly expensive energy. Under any situation, the energy must be conserved, keeping the infrastructure intact. Currently, Saudi Aramco (hence Saudi Arabia) imports these energy-saving value-added flow aids spending a lot of money. Therefore, producing DRP locally is a subject of strategic research importance. Commercial DRPs are made catalytically mostly by polymerizing alpha-olefins. SABIC produces these alpha-olefins at Almuttahida (Jubail United Petrochemical Company) whereas Saudi Organometallic Chemicals Company (SOCC Al-Jubail) makes the related alkyl aluminum cocatalyst. Therefore, to develop an indigenous technology to produce DRP flow aids using the locally available SABIC alpha-olefins and SOCC cocatalyst is timely and appropriate.

The alpha-olefin production capacity of Almuttahida is 150 kmt/yr. SOCC is a joint venture between Saudi Specialty Chemicals Company (SSCC) and Albemarle Netherlands BV. The SOCC alkylaluminum production capacity is 6 kmt/yr. Both Almuttahida and SSCC are SABIC affiliates.

2. Value creation, business development, and macroeconomic matrix

This research, on a national basis, aims at developing Saudi Arabia's economy through conversion of local hydrocarbon raw materials into highly value-added products. The focus is on the following:

i. Maximize value captured from the energy sector;

ii. Unlock potential of non-oil sectors;

iii. Further integrate Saudi economy regionally and globally; and

iv. Grow non-oil exports.

Currently, this study converts SABIC's selected alpha-olefins into value-added drag reducing polymers (DRPs) while alkyl aluminum cocatalyst(s) manufactured by SOCC are planned to be used in the future. Therefore, the above research very well aligns with National Transformation Program (NTP) 2020 and Vision 2030. This alignment defines the niche for value creation.

The DRPs show to be at least 10 times costlier than the starting alpha-olefins. For example, the unit price of a representative alpha-olefin is 0.86 USD/kg whereas that of a related DRP is 9.42 USD/kg. Several global business development companies have published detailed market research reports on DRP. These reports predict steady growth of DRP business worldwide, including Saudi Arabia, till 2023. In this regard, growing crude oil, refinery product, and petroleum industry are the major market drivers. Particularly, the crude oil flow improver (including mostly DRP) global market is estimated to grow from USD 1.282 billion in 2016 to USD 1.920 billion, at a CAGR (compound annual growth rate) of 5.7% during 2017 to 2023. Here, Middle East, Africa, and Asia Pacific are the market leaders because the maximum crude oil extraction comes from Saudi Arabia and Nigeria, followed by the USA, Canada, Europe, and Latin America.

The alpha-olefin feedstock and the experimental cocatalyst are locally available. Therefore, the process and end-products (to be developed) will be very cost-effective, which is an economic benefit. This research initiative, in parallel, holds a great potential to (i) open spin-off business (by non-governmental agencies) in the area of drag reducing flow aids, (ii) reduce Saudi Arabia's dependence on the very costly imported products, and (iii) add to the gross domestic product (GDP). The anticipated social benefit includes enhancement of job creation and better Saudi employment situation. Successful Saudi drag reducing polymers (DRPs) can be easily exported at least, and to begin with, to the neighboring oil-producing countries. Therefore, the trade balance—the difference between Saudi Arabia's imports and exports—is very likely to be favored. Accordingly, prospective local and neighborhood markets exist. Therefore, the proposed research initiative holds very high potential for generating revenue.

The stake holders (beneficiaries) of the proposed initiative include Saudi Aramco, SABIC, key players (such as Conoco Phillips, Baker Hughes, Flowchem, Innospec, Oil Flux Americas, Sino Oil King Shine Chemical, Superchem Technology, and Zoranoc Oilfield Chemical), as well as prospective local specialty chemical business company such as SOCC.

3. Technical review on catalyst development

The literature on drag reducing polymer (DRP) can be mainly divided into journal publications and patents. The journal publications mostly address, from an applied research perspective, the performance evaluation of commercial UHMW DRPs as a function of hydrodynamic variables, conduit configuration, DRP MW and stability, and surfactant type [Wang 2011, Research Institute 2015]. These references can be pursued for the details. However, the research theme of this symposium and our contribution to it is catalyst. Therefore, we review DRP catalysts which are mostly covered by patents. The central concept comprises using the second generation $TiCl_3$ Ziegler-Natta (Z-N) precatalyst with an organoaluminum cocatalyst [Dehm 1983 and Motier 1985].

All the patents modify the above basic catalyst system by introducing a Lewis base (LB) electron donor to block the active centers that catalyze the undesirable formation of low molecular weight polyolefin backbones. The referenced LBs can be divided into Type I LB and Type II LB. Type I is non-phosphorous LBs and Type II is phosphorous LBs. Type I includes ether [Mack et al. 1982], ketone [Mack 1985a; and Mack 1985b], pyridine and pyperidine [Marullo et al. 1964], etc. On the other hand, Type II lists phosphine, phosphite, etc. [Mack 1984]. Additionally, a third family of patents uses aluminoxanes alone, or combined with Z–N alkyl aluminum cocatalysts [Eaton et al. 2000a; Eaton et al. 2000b; Eaton et al. 2001; Eaton et al. 2004a; and Eaton et al. 2004b]. All these patents evidently show variation in components and composition. The measurement of standard solution viscosity mostly characterized the resulting poly(alpha-olefin)s. These data concluded the formation of UHMW DRPs. Despite the fact that the products successfully demonstrated drag reducing performance, the following criticism holds:

- Decrease in catalyst activity;
- Requirements of 0 °C and below, and very long polymerization time (≥ 12 hr);
- Low drag reduction (13.4%);
- Use of hazardous catalyst adjuvant(s), and much costlier gel-prone aluminoxane cocatalysts; and
- Absence of catalyst characterization, fundamentals, and insight.

Therefore, enough room to develop high-performance catalyst and DRP polymerization process exists.

4. Our patented catalyst design concept and composition

The organometallic-catalyzed synthesis of UHMW DRP has two requirements. One is kinetic and the other is thermodynamic. The kinetic requirement demands that the rate of chain propagation R_p should be, on principle, highly large. On the contrary, the rate of chain termination R_t (originating from various active chain transfer reactions) should approach zero. The thermodynamic requirement, in terms of activation energy, is just the opposite. See Figure 2. An embedded challenge in this context is the following. In organometallic-catalyzed olefin polymerization, the catalyst activity and the resulting MW are inversely related. How to establish a balance between both poses a challenge. Our patented catalyst design concept attempts meeting the above requirements and overcoming this challenge.

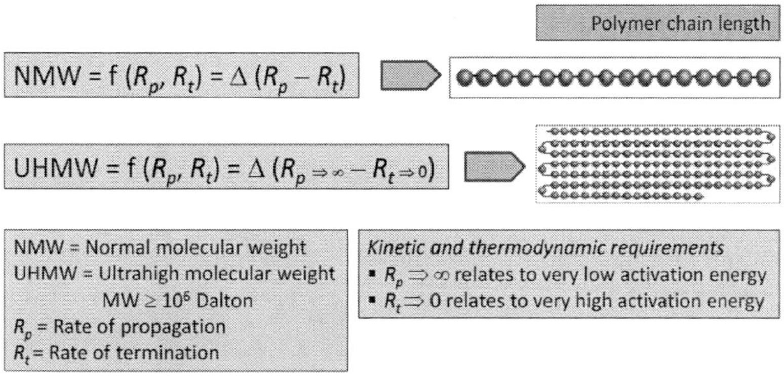

Figure 2. Kinetic and thermodynamic requirements for producing UHMW DRP.

The above catalyst design concept is summarized here. To meet Figure 2 kinetic and thermodynamic requirements, we propose to use Lewis base electron donors that feature the following:

- They simultaneously show steric hindrance and electronic effect. Steric hindrance is achieved by the bulky substituents around the heteroatom which reduce the access of the related base functionality to the Lewis acid alkyl aluminum cocatalyst.

- Electronic effect is obtained by placing electron-withdrawing substituents on the heteroatom to reduce the electron density on the basic heteroatom. Consequently, the alkyl aluminum cocatalyst•Lewis base complex presumably dissociates under the DRP synthesis condition, and consequently the catalyst productivity increases. Aromatic substituents are especially useful because they are relatively unreactive toward other catalyst components.

Our patents [Atiqullah et al. 2017; and Atiqullah et al. 2018] illustrate the above overall catalyst constitutive coordination/complexation concept and structure first by forming an alkylaluminum cocatalyst•Lewis base (LB) complex, and next contacting this complex with the transition metal precatalyst to make the final catalyst. A preferred example of the above LB is a tertiary monophenyl amine, belonging to the group comprising N,N-diethylaniline, N-ethyl-N-methylparatolylamine, N,N-dipropylaniline, N,N-diethylmesitylamine, and combinations thereof.

5. Results and discussion

All the procedures, related to catalyst formulation, DRP synthesis, and drag reducing (DR) performance evaluation, are detailed in our patents [Atiqullah et al. 2017; and Atiqullah et al. 2018]. Therefore, they are only briefed here (as appropriate). A local kerosene was used as the transportation fluid.

The DRPs in this work refer to 1-hexene–1-dodecene copolymers. Almuttahida donated 1-hexene and 1-dodecene. These DRPs were synthesized using the reactor system shown by Figure 3. Figure 4 displays some representative DRP samples. The DR performance was evaluated using the DRP test facility designed, installed, and commissioned by Prof. Abdelsalam Al-Sarkhi, Department of Mechanical Engineering, KFUPM. AP Miniplant Germany fabricated this equipment. See Figures 5 and 6.

Figure 3 Poly(alpha-olefin) DRPs under synthesis.

Figure 4. Selected as-synthesized poly(alpha-olefin) DRPs.

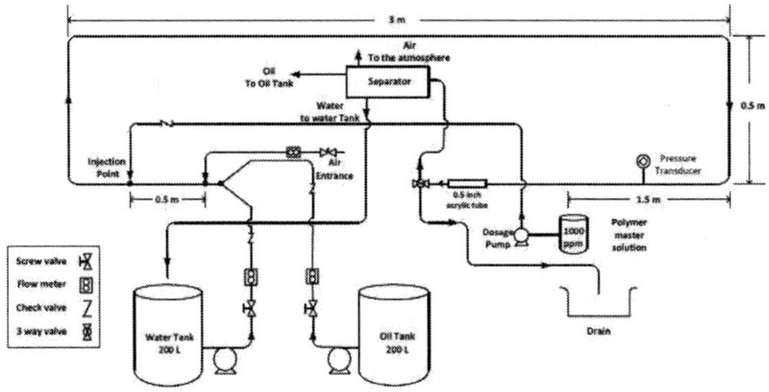

Figure 5. DRP test facility at CRP RI, KFUPM.

Figure 6. The DRP performance test loop under operation.

Before DRP performance evaluation, the test loop was calibrated. See Figures 8 and 9. Experimental and calculated ΔP/Ls (pressure gradients) match very well.

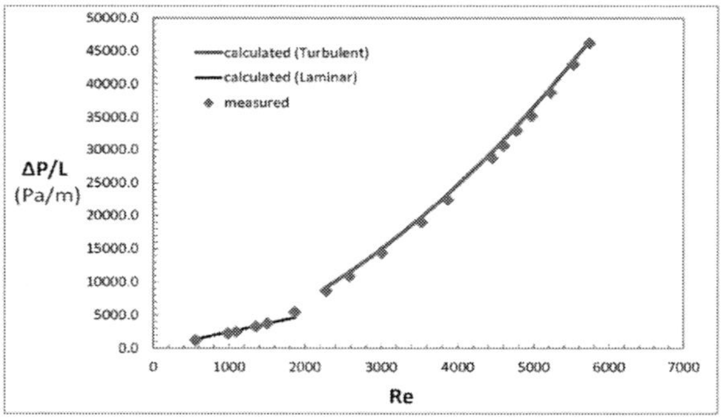

Figure 8. Variation of ΔP/Ls (pressure gradients) with Reynolds number for crude oil flow in a 0.4 inch diameter horizontal pipe in the loop.

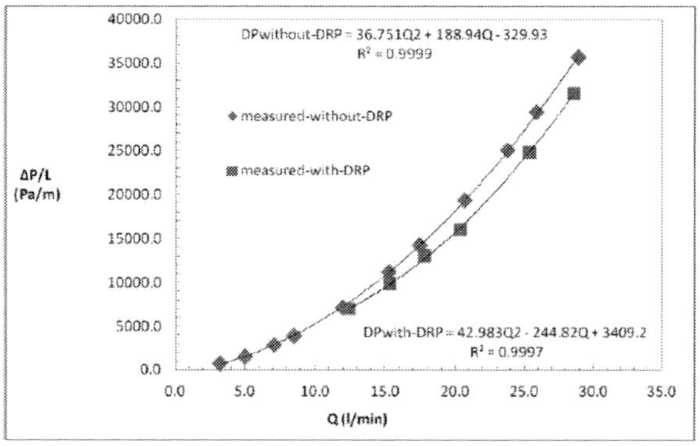

Figure 9. ΔP/L versus volumetric flow rate Q for UHMW polyisobutene (PIB) (a commercial DRA) for 50 ppm concentration in the pipe line.

The %DR was determined using the following pressure drop relation applicable to a horizontal liquid transportation pipe [Subhash et al., 2006; Al-Wahaibi et al., 2007; and Li et al., 2011]:

$$\% \text{ Drag reduction} = \frac{\Delta P_{PF} - \Delta P_{RDF}}{\Delta P_{PF}} \times 100 \qquad (1)$$

where ΔP_{RDF} and ΔP_{PF} are the pressure drops (corresponding to the same flow rate) over a given length of pipe for the experimental fluid with and without the DRP, respectively.

Figure 10 compares the productivity of a given example of the patented catalyst with that of the reference catalyst. The patented catalyst formulation, prepared as per the novel concept reported in Section 4, shows to be 1.4 times more productive than the reference catalyst. Note that increased catalyst productivity increases production volume, which consequently reduces production cost. Therefore, our patented catalyst adds value to DRP process development and economics.

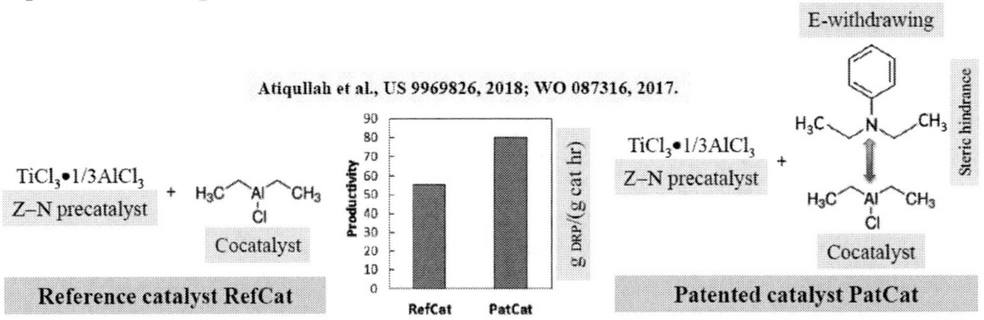

Figure 10. Comparison of productivity of reference versus patented catalyst.

Figure 11 shows the effects of catalyst formulation type, and DRP MW and microstructure on DR%. By DRP microstructure, we mean monomer/comonomer sequence and side chain distribution in the copolymer backbone. Several important findings are evident. These are summarized below.

First, compare the DR% of PatCat DRP with that of RefCat DRP. The former approaches that of the latter even for a much lower MW value. The MW of PatCat DRP is 1.70 million Dalton while that of RefCat DRP is 2.37 million Dalton. This means that the patented catalyst is capable of regulating the microstructure of the 1-hexene–1-dodecene copolymer in such a manner that commensurate DR% can also be achieved using lower MW value DRP with higher activity. Increased UHMW DRP has been reported in the literature as pre-requisite to improving DR%. This work does not favor this argument. To the best of our knowledge, such a finding has not been reported in the literature. This finding has significant scientific value. This will motivate the concerned researchers to revisit the current mechanism of turbulent drag reduction and DRP rheology which will, in turn, set new directions to fundamental and applied research of this important subject. The 45% DR of the PatCat DRP signifies the following. The resulting product almost triples the rate of transportation of a selected grade of refinery product with a saving of about 50% pumping energy at ppm level pipeline concentration [Lescarboura et al. 1971].

Additionally, the use of increased UHMW DRPs has the following practical disadvantages. They dissolve less readily in hydrocarbon fluid than the lower UHMW value DRPs. The solution viscosity also increases with MW. Therefore, the injection procedure becomes cumbersome and more energy-consuming. Therefore, we show how these disadvantages can be overcome through designing the catalyst in an appropriately novel fashion.

Second, evaluate the DR% performance of PatCat DRP against that of CP LP100 and SP PIB. CP LP100 is a poly(1-decene)–based Conoco Phillips DRP formulation whereas SP PIB is an UHMW poly(isobutene) from Scientific Polymer Products Inc. Both products are homopolymers. This evaluation also establishes that the as-synthesized PatCat DRP, despite having lower UHMW, shows comparable to higher DR% without making a DRP formulation. This is another manifest of the merits of the patented catalyst and the 1-hexene–1-dodecene DRP it synthesizes. This improved PatCat DRP performance may be attributed to the potential enhancement of the extensional viscosity of the above 1-hexene–1-dodecene copolymer. The present catalyst-mediated copolymerization route is apt to shift the DRP synthesis process from 0 °C and below to room temperature and above.

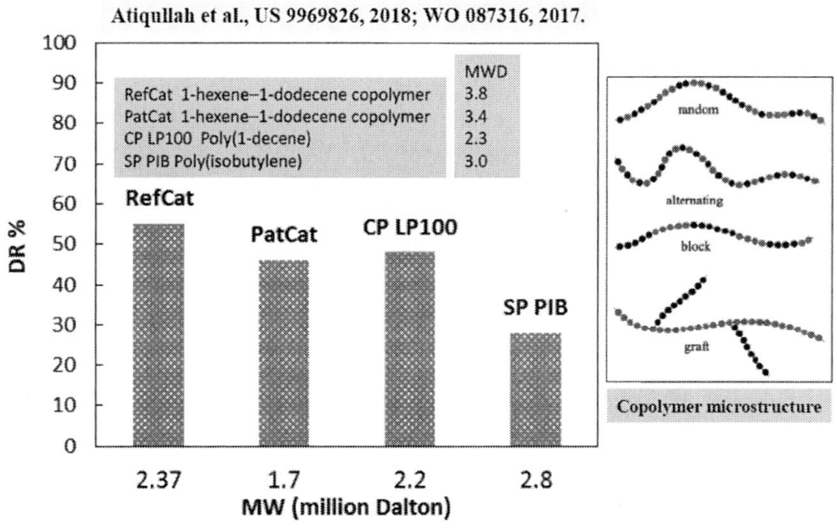

Figure 11. Effect of catalyst formulation type, MW, and DRP microstructure on DR%.

6. Future challenges and research niches

This subject includes future challenges and research niches in DRP catalyst and process development, product formulation development, and DRP injection into the pipeline. These are categorically summarized below.

6.1 DRP catalyst and process development

The developments of high-performance DRP catalysts and processes are inter-related. The current process and catalyst mostly synthesize a single-phase solution product. They require very low temperature and lengthy polymerization time. Undesirable rod-climbing effect occurs with progression of polymerization due to stirring as viscosity develops and normal stress exceeds tangential stress. This adversely affects macro- and micromixing of the highly viscous reaction mixture. Consequently, consistency in product quality deteriorates. The resulting product dissolves in the alpha-olefin comonomers and the solvent used. This prevents achieving in-situ particulate morphology (during polymerization) which is a major catalyst-cum-process development challenge. Such a morphology can make the DRP ready for transportation and direct use without significant post-synthesis product formulation.

Comprehensive investigation of catalyst-cocatalyst-donor complexation and dissociation behavior, and product microstructural characterization can further elucidate the catalyst and DRP process performance. However, these subjects show to be unfortunately undone and/or unpublished.

6.2 Product formulation development

The product development challenges list the following:

- Pumpable stable slurry without viscofication;
- Elimination of phase separation instability;
- Agglomeration and gel formation;
- Transportation inconvenience;
- Prompt DRP dissolution within turbulent mixing length; and
- Achievement of very high extensional viscosity.

The currently used cryogenic and grinding process—a prerequisite to DRP formulation—is very cumbersome and costly. This requires maintaining the storage and transportation temperature above certain critical value. Also, heated injectors are to be often used to introduce the DRP into the transportation line. Dipropylene glycol monomethyl ether (DPGME) is widely used. However, this oxidizes readily in air to form unstable peroxides that may explode spontaneously. Most seriously, cryogenic grinding destructs the UHMW polymer backbones. Consequently, the DRP performance significantly decreases.

6.3 DRP injection into the pipeline

The DRP pipeline injection problems include the following:

- Low ppm level injection;
- Consistent pumping and feeding; and
- Elimination of turbulence- and shear-induced mechanical degradation.

Modeling efforts that can correlate turbulent pipe flow hydrodynamics to DRP catalyst and process as well as product development are fairly inadequate.

7. Our vision

Our vision is to develop an interdisciplinary world-class DRP research program at KFUPM through collaboration among the Center for Refining & Petrochemicals (CRP), Departments of Mechanical Engineering, Chemistry, and Chemical Engineering (as appropriate), Saudi Aramco, SABIC, SOCC, Ministry of Education, and a prospective international collaborator. KFUPM has the relevant human and material resources, experience, and expertise in this area to move ahead. We have creative ideas and novel approaches to address the challenges summarized in Section 6. Achieving this vision will serve a great national cause.

8. Conclusions

The background and importance of this research, which aligns with Vision 2030, have been underscored from the perspective of Saudi Aramco, SABIC, and SOCC. Patents, unlike the journal publications, cover the catalytic synthesis of poly(alpha-olefin) drag reducing polymers. Therefore, they have been thoroughly reviewed. This review concludes that the patents unfortunately do not address the catalytic science. The merits of the present patented catalytic system have been summarized. The premier contribution is the following. It can regulate the microstructure of the 1-hexene–1-dodecene copolymer in such a manner that lower MW value UHMW product can also achieve commensurate DR% with higher activity. This finding will set new direction to researching turbulent drag reduction mechanism and DRP rheology. The present catalyst-mediated copolymerization route shows promise to shift the DRP synthesis process from 0 °C and below to room temperature and above. Future research niches and challenges have been outlined. Our vision to develop an interdisciplinary world-class DRP research program at KFUPM has been presented.

Acknowledgements

We highly appreciate Saudi Aramco for supporting this research through PN CRP02253 under University Collaboration Program (UCP). The donation of selected higher alpha-olefins (1-hexene, 1-octene, 1-decene, and 1-dodecene) by Almuttahida SABIC is also thankfully acknowledged. The assistance of Mr. Jaber Al-Qahtani is well appreciated.

References

Al-Roomi Y, George R, Elgibaly A, and Elkamel A. Use of a novel surfactant for improving the transportability/transportation of heavy/viscous crude oils. *Journal of Petroleum Science and Engineering*, v. 42, 235–243 (2004).

Al-Surakji, IHH, Alsarkhi AMH, Atiqullah M, El-Nakla M, and Alhems, LM. Study of oil-soluble and water-soluble drag reducing polymers in multiphase flow. *Canadian Journal of Chemical Engineering*. v. 9999, 1–17 (2017).

Al-Wahaibi T, Smith M, and Angeli P. Effect of drag-reducing polymers on horizontal oil–water flows. *Journal of Petroleum Science and Engineering*, v. 57, 334–346 (2007).

Atiqullah M, Al-Sarkhi AS, Al-Thenayan FM, Al-Malki AR, Xu W, and Hossaen A. Catalyst composition and a process for making ultrahigh molecular weight poly(alphaolefin) drag reducing agents. US Patent 9,969,826 B2, May 15 (2018).

Atiqullah M, Al-Sarkhi AS, Al-Thenayan FM, Al-Malki AR, Xu W, and Hossaen A. Catalyst composition and a process for making ultrahigh molecular weight poly(alphaolefin) drag reducing agents. WO/2017/087316, May 26 (2017).

Dehm DC. Low temperature polymerization process. US Patent 4,384,089 (1983).

Eaton GB, Monahan MJ, Tipton and RJ. Methods for forming amorphous ultra-high molecular weight polyolefins and drag reducing compositions comprising amorphous ultra-high molecular weight polyolefins. USP 6,730,752 (2004a).

Eaton GB, Monahan MJ, Ebert AK, Tipton RJ, and Baralt E. Methods for forming amorphous ultra-high molecular weight polyolefins for use as drag reducing agents. USP 6,730,750 B2 (2004b).

Eaton GB, Monahan MJ, and Tipton RJ. Methods for forming amorphous ultra-high molecular weight polyalphaolefin drag reducing agents. USP 6,015,779 (2000a).

Eaton GB, Monahan MJ, Tipton RJ. Methods for forming amorphous ultra-high molecular weight polyalphaolefin drag reducing agents using a halocarbon. USP 6,162,773 (2000b).

Eaton GB, Monahan MJ, and Tipton RJ. Methods for forming amorphous ultra-high molecular weight polyalphaolefin drag reducing agents using non-metallocene and alkylaluminoxane. USP 6,242,395 (2001).

Hasan SW, Ghannam MT, and Esmail N. Heavy crude oil viscosity reduction and rheology for pipeline transportation. *Fuel*, v. 89, 1095–1100 (2010).

Lescarboura JA, Culter JD, and Wahl HA. Drag Reduction with a Polymeric Additive in Crude Oil Pipelines. *Society of Petroleum Engineering*, v. 1. 229–234 (1971).

Li B, Dong G, and Zhang C. Storage stability and solubility of poly(urea-formaldehyde) microcapsules containing α-olefin drag reducing polymer. *Journal of Applied Polymer Science*, v. 122, 1450–1456 (2011).

Mack PM, Decker LB, and Wallac AL. Method for the preparation of non-crystalline polymers of high molecular weight. USP 4,358,572 (1982).

Mack PM. Catalyst and method for preparation of drag reducing substances. USP 4433123 A (1985b).

Mack PM. Polymerization process for drag reducing substances. USP 4,433,123 (1984).

Mack PM. Polymerization process for drag reducing substances. USP 4493903 A (1985a).

Marullo G, Baroni A, and Maffezzoni U. Three component catalyst for polymerization of 1-olefins containing titanium halide, aluminum alkyl, and pyridine. USP 3,139,418 (1964).

Martĭnez-Palou R, Mosqueira MDL, Zapata-Rendōn B, Mar-Juárez E, Bernal-Huicochea C, Clavel-López JdlC, and Aburto J. Transportation of heavy and extra-heavy crude oil by pipeline: A review. *Journal of Petroleum Science and Engineering*, v. 75, 274–282 (2011).

Motier J. Compositions for and method of reducing hydrocarbon fluid friction loss in conduits. USP 4,527,581 (1985).

Mowla D and Naderi A. Experimental study of drag reduction by a polymeric additive in slug two-phase flow of crude oil and air in horizontal pipes. *Chemical Engineering Science*, v. 61, 1549–1554 (2006).

Research Institute. Development of Novel Poly(alpha-olefin) Drag Reducing Agents for Saudi Aramco oil and Refinery Products. Final Report CRP02253, October (2015).

Subhash NS, Ahmed K, and Yunxu Z. Drag reduction characteristics in straight and coiled tubing—An experimental study. *Journal of Petroleum Science and Engineering*,
v. 53 179–188 (2006).

Toms BA. Some observations on the flow of linear polymer solutions through straight tubes at large Reynolds numbers. *Proceedings of the 1st International Congress on Rheology*, North Holland, v. 2, 135–141 (1949).

Wang Y, Yu B, Zakin JL, and Shi H. Review on drag reduction and its heat transfer by additives. *Advances in Mechanical Engineering*, Volume 2011, Article ID 478749, 17 pages doi:10.1155/2011/478749.

PAPER 24

Novel cycloolefin copolymers synthesized by *ansa*-Dimethylsilylene (fluorenyl) (amido) titanium-based catalysts

Takeshi Shiono
Graduate School of Engineering, Hiroshima University
1-4-1 Kagamiyama, Higashi-Hiroshima 739-8527, Japan
TEL: 81-82-424-7730, FAX: 81-82-424-5494, e-mail: tshiono@hiroshima-u.ac.jp

Cyclic olefin copolymers (COCs) which consists of rigid alicyclic polymer-backbone are attractive materials because of their good heat and chemical resistance as well as their low dielectric constants, nonhygroscopicity, and high transparency. Despite the expectation for copolymerization of cyclic olefins with higher 1-alkene giving new COCs, the catalytic systems available for the copolymerization are limited and their activity and the molecular weights of the products were not sufficient.

We have fortunately found that *ansa*-dimethylsilylene(fluorenyl)(*t*-butylamido) dimethyl titanium complexes (**1**) conducted living polymerization of norbornene and 1-alkene activated by a suitable cocatalyst. Excellent copolymerization ability of **1** also enables us to introduce functional groups in COC via copolymerization of norbornene with functionalized 1-alkenes as shown in Figure 1. In this presentation, I introduce novel COCs synthesized by the use of **1**-based catalytic systems.

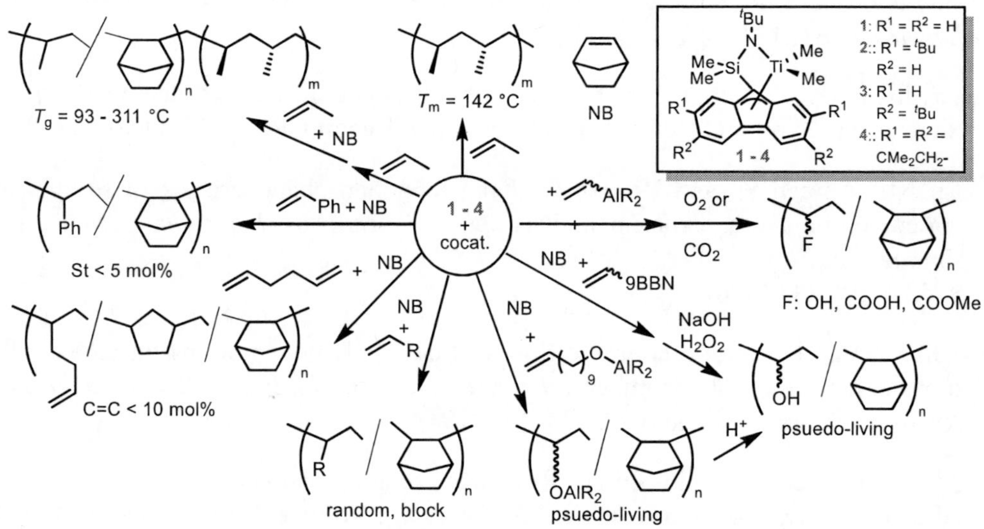

Figure 1. Novel norbornene copolymers synthesized by *ansa*-dimethylsilylene(fluorenyl)-(*tert*-butylamido)dimethyltitanium-based catalysts

References

[1] Zhao, W.; Nomura, K. *Macromolecules* 2016, *49*, 59.

[2] He, X.; Deng, Y.; Han, Z.; Yang, Y.; Chen, D. *J. Polym. Sci., Part A: Polym. Chem.* 2016, *54*, 3495.

[3] Shiono, T.; Sugimoto, M.; Hasan, T.; Cai, Z.; Ikeda, T. *Macromolecules* 2008, *41*, 8292.

[4] Cai, Z.; Harada, R.; Nakayama, Y.; Shiono, T. *Macromolecules* 2010, *43*, 4527.

[5] Ban, H. T.; Hagihara, H.; Nishii, K.; Tsunogae, Y.; Nojima, S.; Shiono, T. *J. Polym. Sci., Part A: Polym. Chem.* 2008, *46*, 7395: Ban, H. T.; Hagihara, H.; Tsunogae, Y.; Cai, Z.; Shiono, T. *ibid.* 2011, *49*, 65.

[6] Tanaka, R.; Matsuzaki, R.; Nakayama, Y.; Shiono, T. *J. Polym. Sci. Part A: Polym. Chem.* 2017, *55*, 2136: Tanaka, R.; Sasaki, A.; Takenaka, T.; Nakayama, Y.; Shiono, T. *Polymer* 2018, *136*, 109.

[7] Lee, J.-W.; Jantasee, S.; Jongsomsjit, B.; Tanaka, R.; Nakayama, Y.; Shiono, T. *J. Polym. Sci., Part A: Polym. Chem.* 2013, *51*, 5085: Tanaka, R.; Suenaga, T.; Cai, Z.; Nakayama, Y.; Shiono, T. *ibid.* 2014, *52*, 267:Tanaka, R.; Nakayama, Y.; Shiono, T. *Polym. Chem.* 2013, *4*, 3974: Tanaka, R.; Goda, M.; Cai, Z.; Nakayama, Y.; Shiono, T. *Polymer* 2015, *70*, 252: Tanaka, R.; Ikeda, T.; Nakayama, Y.; Shiono, T. *ibid.* 2015, *56*, 218: Song, X.; Yu, L.; Shiono, T.; Hasan, T.; Cai, Z. *Macromol. Rapid Commun.* 2017, *38*, 1600815: Tanaka, R.; Tonoko, N.; Kihara, S.-i.; Nakayama, Y.; Shiono, T. *Polym. Chem.* 2018, *9*, 377.

28th Annual Saudi-Japan Symposium: Nov. 12-13, 2018

Technology in Petroleum Refining & Petrochemicals

Venue: KFUPM Dhahran Techno Valley, Innovation Cluster Bldg., Auditorium 1111

Day One: Monday, November 12, 2018

OPENING REMARKS SESSION — *Chairman: Dr. Hassan S. Al-Asiri*

8:15 *Registration & Coffee Break*
8:30 *Opening Remarks*

- *H.E. Dr. Sahel Abuduljauwad, Rector, KFUPM*
- *Mr. Ahmed Al-Khuwaiter, CTO, Saudi Aramco*
- *Mr. Eiji Hiraoka, Senior Executive Director, JCCP*
- *Dr. Koichi Eguchi, President, JPI*

SESSION ONE CLEAN FUELS AND HYDROGEN — *Chairman: Dr. Takeshi Shiono*

9:00	1.	**Keynote: Characteristics of hydrogen carriers and application to fuel cells,** *Dr. Koichi Eguchi, JPI*
9:30	2.	**Hydrogen production through thermo-neutral reforming of liquid fuels using structured catalysts,** *Dr. Shakeel Ahmed, KFUPM/CRP*
9:50	3.	**Catalytic steam reforming of biomass-derivatives for hydrogen production,** *Dr. Yasushi Sekine, Waseda University, Japan*
10:10	4.	**Design of scalable and stable catalysts for CO_2 reforming of methane,** *Dr. Oki Muraza, KFUPM/CENT*
10:30		*Coffee Break*

SESSION TWO NEW CHALLENGES: FUELS/CHEMICALS — *Chairman: Dr. Ali H. Zaid*

10:50	5.	**Keynote: Future fuels & petrochemicals: new technologies in the face of new challenges** *Dr. James Johnson, Honeywell UOP, USA*
11:20	6.	**Clean energy generation by artificial photosynthesis based on semiconductor technology** *Dr. Kazuhiro Ohkawa, KAUST*
11:40	7.	**The effect of lanthanum exchange on USY-zeolites for the alkylation of 2-butene with isobutane,** *Dr. Mohammed Al-Dossary, Saudi Aramco R&DC*
12:00		*Prayer & Lunch Break*

SESSION THREE IMO REGULATIONS/HEAVY OIL — *Chairman: Dr. Khalid Al-Hooshani*

13:00	8.	**Responding to 2020 IMO SOx regulation,** *Mr. Atsunori Sato, JGC Corporation, Japan*
13:20	9.	**Effect of zeolite Y modification on reaction performance of heavy crude oil hydrocracking,** *Dr. Lianhui Ding, Saudi Aramco R&DC*
13:40	10.	**Oil soluble dispersed catalysts for slurry phase hydrocracking of heavy VGO,** *Mr. Tareq Al-Attas, KFUPM/ChE*
14:00	11.	**Development of *Petroleomics* from fundamentals to applications,** *Dr. Kazuhiro Inamura, Japan Petroleum Energy Center (JPEC), Japan*
14:20		*Day One Ends*

Each presentation includes 5-minutes Q&A — http://catsymp.kfupm.edu.sa/

28th Annual Saudi-Japan Symposium: Nov. 12-13, 2018

Technology in Petroleum Refining & Petrochemicals

▲▲▲ *Venue: KFUPM Dhahran Techno Valley, Innovation Cluster Bldg., Auditorium 1111* ▲▲▲

Day Two: Tuesday, November 13, 2018

SESSION FOUR CATALYST TECHNOLOGY/AROMATICS ***Chairman: Dr. Yasushi Sekine***

8:00 *Registration & Coffee Break*

8:30 12. **Effect of alumina binder on deactivation behavior of Pt-Beta catalyst for transalkylation of toluene with 1,2,4-trimethylbenzene,** *Dr. Faisal Al-Mulla, Saudi Aramco R&DC*

8:50 13. **Driving catalysis innovation in petrochemicals through designing of unneglectable support materials,** *Dr. Muhammad Haider, SABIC Riyadh*

9:10 14. **Study the nano-structured zeolites for hydrocarbon reforming: kinetic and reaction pathways,** *Mr. Ali N. Al-Jishi, Saudi Aramco R&DC*

9:30 15. **Advance characterization of heterogeneous catalyst of core-shell Pt@SiO2 nanoparticles,** *Dr. Noor Al-Mana, Saudi Aramco R&DC*

9:50 *Coffee Break*

SESSION FIVE PETROCHEMICAL APPLICATIONS ***Chairman: Mr. Alhussain Bakhurji***

10:20 16. **Numerical modeling for selected chemical processes,** *Dr. Tarek Jamaleddine, SABIC Riyadh*

10:40 17. **Oxidative reforming of n-butane to ethylene, propylene and syngas on acid/base nano-hybrid catalyst,** *Dr. Sachio Asaoka, KFUPM/CRP and JCCP*

11:00 18. **Light naphtha upgrading to aromatics using metal modified MFI zeolite catalysts,** *Dr. Yaming Jin, Saudi Aramco R&DC*

11:20 19. **Upgrading of unused heavy fuel oil over iron oxide based catalysts under sub and supercritical water conditions,** *Dr. Takuya Yoshikawa, Hokkaido University, Japan*

11:40 20. **Zeolite catalyst development for enhancing p-xylene selectivity,** *Dr. Mohammed Al-Bahar, Saudi Aramco R&DC*

12:00 *Prayer & Lunch Break*

SESSION SIX COMPOSITES AND POLYMERS ***Chairman: Dr. Mohammed Ba-Shammakh***

13:00 21. **MgO-blended polyethylene glycol composites as a shape-stabilized phase change material for solar thermal energy storage,** *Dr. Md. Hasan Zahir, KFUPM/CoRE-RE*

13:20 22. **Nuclear Magnetic Resonance (NMR) Spectroscopy at Saudi Aramco's R&D Center: Where Synergy is Opportunity,** *Dr. Qasim Saleem, Saudi Aramco R&DC*

13:40 23. **Catalytic conversion of Saudi alpha-olefins to energy-saving value-added polymeric flow aids,** *Dr. Muhammad Atiqullah, KFUPM/CRP*

14:00 24. **Novel cycloolefin copolymers synthesized by ansa-dimethylsilylene fluorenyl)(amido) titanium-based catalysts,** *Dr. Takeshi Shiono, Hiroshima University, Japan*

14:20 ***Closing Remarks, Symposium Ends***

Each presentation includes 5-minutes Q&A *Program as of Oct. 22, 2018*

Registration: http://catsymp.kfupm.edu.sa/register.htm
Center for Refining & Petrochemicals: Phone 013 860-2029